Charlotte Henkel

An Algorithm for Linear Stochastic Bilevel Problems

Charlotte Henkel

An Algorithm for Linear Stochastic Bilevel Problems

Südwestdeutscher Verlag für Hochschulschriften

Impressum / Imprint
Bibliografische Information der Deutschen Nationalbibliothek: Die Deutsche Nationalbibliothek verzeichnet diese Publikation in der Deutschen Nationalbibliografie; detaillierte bibliografische Daten sind im Internet über http://dnb.d-nb.de abrufbar.
Alle in diesem Buch genannten Marken und Produktnamen unterliegen warenzeichen-, marken- oder patentrechtlichem Schutz bzw. sind Warenzeichen oder eingetragene Warenzeichen der jeweiligen Inhaber. Die Wiedergabe von Marken, Produktnamen, Gebrauchsnamen, Handelsnamen, Warenbezeichnungen u.s.w. in diesem Werk berechtigt auch ohne besondere Kennzeichnung nicht zu der Annahme, dass solche Namen im Sinne der Warenzeichen- und Markenschutzgesetzgebung als frei zu betrachten wären und daher von jedermann benutzt werden dürften.

Bibliographic information published by the Deutsche Nationalbibliothek: The Deutsche Nationalbibliothek lists this publication in the Deutsche Nationalbibliografie; detailed bibliographic data are available in the Internet at http://dnb.d-nb.de.
Any brand names and product names mentioned in this book are subject to trademark, brand or patent protection and are trademarks or registered trademarks of their respective holders. The use of brand names, product names, common names, trade names, product descriptions etc. even without a particular marking in this work is in no way to be construed to mean that such names may be regarded as unrestricted in respect of trademark and brand protection legislation and could thus be used by anyone.

Coverbild / Cover image: www.ingimage.com

Verlag / Publisher:
Südwestdeutscher Verlag für Hochschulschriften
ist ein Imprint der / is a trademark of
OmniScriptum GmbH & Co. KG
Heinrich-Böcking-Str. 6-8, 66121 Saarbrücken, Deutschland / Germany
Email: info@svh-verlag.de

Herstellung: siehe letzte Seite /
Printed at: see last page
ISBN: 978-3-8381-5037-6

Zugl. / Approved by: Duisburg, Universität Duisburg-Essen, Dissertation, 2014

Copyright © 2015 OmniScriptum GmbH & Co. KG
Alle Rechte vorbehalten. / All rights reserved. Saarbrücken 2015

Abstract

The aim of this thesis is to find a technique that allows for the use of decomposition methods known from stochastic programming in the framework of linear stochastic bilevel problems. The uncertainty is modeled as a discrete, finite distribution on some probability space. Two approaches are made, one using the optimal value function of the lower level, whereas the second technique uses the Karush-Kuhn-Tucker conditions of the lower level. Using the latter approach, an integer-programming based algorithm for the global resolution of these problems is presented and evaluated.

Contents

List of Figures	v
1 Introduction	**1**
1.1 Linear Two-Stage Problems – A Special Case	5
1.2 Basic Complexities	6
1.3 Motivation	9
1.4 Literature	10
2 Linear Two-Stage Problems	**17**
2.1 Properties	20
2.2 Decomposition Methods	21
3 Linear Stochastic Bilevel Problems	**25**
3.1 Linear Bilevel Problems	25
3.1.1 Existence of Solutions	33
3.1.2 Reformulations	37
3.2 Under Uncertainty	40
4 Optimal Value Function Approach	**43**
4.1 A Method for Linear Bilevel Programs	47
4.2 Alternatives	52
5 Karush-Kuhn-Tucker Approach	**53**
5.1 Properties in the General Case	53
5.2 Properties in the Linear Case	56

CONTENTS

6 Stolibi – An Algorithm for Linear Stochastic Bilevel Problems   **61**
   6.1 Preliminaries 62
   6.2 Decomposition 67
   6.3 Coverage of all Cases 70
   6.4 Search on the Binary Set 72
   6.5 Pseudo-Code of Stolibi 78
       6.5.1 Preprocessing 78
       6.5.2 Main Procedure 81
       6.5.3 Sparsification Procedure 85
   6.6 Modulations of the Original Algorithm 90
   6.7 Proof of Correctness and Finiteness 90
   6.8 Requirements and Input 93
   6.9 Computational Results 94

7 Conclusion   **99**

References   **101**

List of Figures

1.1 Possible form of a pessimistic bilevel value function, found in Dempe [28].   9

2.1 Dual block-angular structure, see Birge and Louveaux [15]. 21
2.2 Block-angular structure, also called L-shaped, see Birge and Louveaux [15]. 23

3.1 The feasible set of the bilevel problem is only described by the two thick-lined edges connecting A and E, E and D, respectively, found in Dempe [28]. 26
3.2 Possible form of a pessimistic bilevel value function, found in Dempe [28]. 29

6.1 Flow chart of the preprocessing. 80
6.2 Flow chart of the main procedure. 84
6.3 Flow chart of the sparsification procedure. 89

LIST OF FIGURES

1

Introduction

Bilevel programming problems are hierarchical mathematical optimization problems between at least two groups of decision makers. Each group has a decision vector that can affect the other group's or groups' constraints and objective functions. Besides, every decision vector is chosen optimally to its own groups optimization problem that can again be influenced by the other decision vectors.

The problem is hierarchical which means that the first group chooses its variables including the optimal reaction of the other groups. So, the second group receives the first group's choice as a parameter and solves its optimization problem. The optimal decision vector of the second group can be seen as a (multi-)function dependent on the first groups choice.

If there are more groups, the hierarchy continues analogously with dependence of the third groups choice on the first and second groups' decisions, and so on.

Mathematically spoken, bilevel optimization problems are mathematical programs that have in addition to their normal constraints – whether linear or not – "a subset of their *(decision)* variables constrained to be an optimal solution of other programs parameterized by the remaining variables. When these other programs are pure mathematical programs we are dealing with bilevel programming", see Vicente and Calamai [89].

They are very closely related to special game theoretical questions, namely Stackelberg games which are named after a work of Heinrich von Stackelberg published in 1934 [90]. The players in this strategic game are acting hierarchically and can choose one strategy out of a finite or infinite pool of strategies defined for each player.

1. INTRODUCTION

The player who acts first is called the *leader* and the player selecting his strategy second is called the *follower*. Again, it is important that the leader has the advantage to move first and the follower can just react to that.

The objective in a Stackelberg game is to find a strategy for the leader that minimizes his costs with respect to both, the leader's strategy and the follower's optimal reaction. The latter implicates that the follower, in turn, chooses the cheapest strategy out of his pool of strategies dependent on the leader's decision.

It is possible that each group of players can contain more than one member. In this case, the members are additionally assumed to search for an equilibrium in between them (e.g. Nash or Stackelberg), see Sherali et al. [82] or Simaan [84].

The difference to bilevel programming lies in the fact that, in bilevel programming, both (re-)action sets can depend on the decision of the other player and that there can be more hierarchies than just two. But, a part of the nomenclature in bilevel programming is adopted from Stackelberg games. So, the first group is also called the leader and the second is the follower.

The applications for this setting are widely spread. It can be employed every time two or more players act and react to each other, examples:

- Decision making for a firm that interacts with a market regulator or the market itself,
- Physical and process design problems,
- Structural optimization problems,
- Optimal control problems, or
- Principal-agency-problems.

The general form of a two-player bilevel program can be described as

$$\min \quad F(x,y)$$
$$\text{s.t.} \quad G(x,y) \leq 0 \quad \quad (1.1)$$
$$y \in Y(x) = \arg\min_{y^*}\{f(x,y^*) : g(x,y^*) \leq 0\}$$

Here, $F: \mathbb{R}^n \times \mathbb{R}^m \to \mathbb{R}$ and $f: \mathbb{R}^n \times \mathbb{R}^m \to \mathbb{R}$ are the upper and lower level objective functions, respectively, as well as $G: \mathbb{R}^n \times \mathbb{R}^m \to \mathbb{R}^{r_1}$ and $g: \mathbb{R}^n \times \mathbb{R}^m \to \mathbb{R}^{r_2}$ the upper and lower level constraint functions, respectively. $Y(x)$ is called the *rational reaction set* of the follower and

$$IR = \{(x,y) \in \mathbb{R}^n \times \mathbb{R}^m : G(x,y) \leq 0, y \in Y(x)\}$$

is called the *inducible* or *induced region*, i.e., the set containing all feasible solutions of that bilevel problem.

It has been shown in many works that this general bilevel problem is in most cases irregular because many of the well-known constraint qualifications from nonlinear optimization do not hold here. Further information about this topic will be given in the corresponding chapters according to its reformulation.

In this thesis, it is assumed that the leader's constraints are independent of the follower's decision, so $G(x,y) = G(x)$. Besides, the problem is assumed to be linear. That means that every objective function and every constraint is linear (or affine) in both decision vectors. Thus, the linear (so far deterministic) model can be written as

$$\min \quad c^\top x + d^\top y$$
$$\text{s.t.} \ Ax = b, \, x \geq 0$$
$$y \in Y(x) = \arg\min_{y^*}\{q^\top y^* : Wy^* + Hx = h, \, y^* \geq 0\}$$

where $x \in \mathbb{R}^n$ is the decision vector of the leader and $y \in \mathbb{R}^m$ that of the follower. The vectors c, d, b, q, and h as well as the matrices A, W, and H are of appropriate sizes.

Now, there can be uncertainties in the follower's problem, either because the leader does not have full information about it or because the problem is dependent on future events that cannot be (easily) predicted. That is when stochasticity comes into play. The field of stochastic programming provides many approaches and methods on how to deal with it. Therefore, a short overview is given in chapter 2 where the reader also finds two decomposition methods that might be useful in stochastic bilevel programming.

The aim of this thesis is to find an algorithm for linear stochastic bilevel problems that allows for the use of decomposition techniques as it is possible in many stochastic programming problems. In particular, two basic approaches relying on different characterizations of optimality are considered, via the optimal value function and via the Karush-Kuhn-Tucker conditions.

In chapter 3, linear bilevel problems are presented as well as some of their properties. In order to track solutions to these problems computationally, three different approaches

1. INTRODUCTION

of reformulation – among others – have been developed and will be discussed along with their relation to the original problem. Additionally, a linear stochastic bilevel problem will be built and further assumptions concerning the stochastics are made. The model of interest has a finite set of outcomes and is formulated such that the upper level has to decide "here-and-now", whereas the lower level is able to choose "wait-and-see".

In the ensuing two chapters, the mentioned reformulation approaches will be examined in detail for the presented linear stochastic bilevel problem. The first unites both feasibility sets into one and adds an additional constraint that uses the optimal value function of the follower. This is done in order to eliminate all those decision variables of the follower that would not be the optimal choice for the follower. In chapter 4, the difficulties are illustrated when aiming at a scenariowise decomposition for reformulations via the optimal value function.

The other significant approach, shown in chapter 5, is to solve the bilevel problem using the Karush-Kuhn-Tucker (abbr. KKT) conditions for the lower level. After the transformation, these problems result in mathematical problems with equilibrium constraints. It is shown that this method is quite advantageous when it comes to decomposition.

An algorithm, named stolibi, that uses the latter approach is displayed in chapter 6. Stolibi is based on the work of Hu et al. [50] where a cutting plane algorithm is proposed for deterministic linear programs with linear complementarity constraints. The algorithm is analyzed and adjusted for the optimistic linear stochastic bilevel problem using the reformulation of the KKT approach, including the use of a decomposition mehtod. The new pseudo-agorithm is displayed as well as an evaluation of numerical results.

Finally, there is a conclusions chapter.

Summarizing, the main contribution of this thesis is the algorithm presented in chapter 6 which allows for the use of a decomposition method. In stochastic programming, especially in two-stage problems, linearity is a well-regarded property that has been deeply studied and used for fast optimization such as decomposition methods. Linear stochastic bilevel problems, though, present a harder class of problems, even in the deterministic case. Therefore, decomposition methods for those problems are still to be established further. Two common approaches to solve bilevel problems are examined for their ability to allow for decomposition. While the first only serves as an example that it might be very hard to find a decomposition method using the optimal value function, an algorithm is found for the KKT approach.

The algorithm is intensely analyzed and proven to work correct. The implemented code serves mostly as a proof of concept and is therefore not comparable to commercial mixed-integer solvers, such as Gurobi [47].

Another contribution can be found in the analysis of linear bilevel problems in section 3.1.

The remaining part of this chapter is organized as follows. The relationship between linear two-stage problems (as known from Birge and Louveaux [15] or Kall and Wallace [57]) and linear stochastic bilevel problems is displayed in section 1.1. It is followed by a section that exhibits the problems that arise in bilevel programming. Section 1.3 then points out the motivation for this work. The chapter concludes with an outline of bilevel and especially stochastic bilevel literature.

1.1 Linear Two-Stage Problems – A Special Case

Linear stochastic bilevel problems and linear two-stage problems bear a lot of similarities: two levels or stages according with two different types of decision variables influencing each other, and both having an important impact on the upper stage or leader's objective. In fact, they both belong to the class of hierarchical planning problems (see Patriksson and Wynter [71]) and so they are very closely related as will be shown now. But still, there exist distinctions that are very crucial.

A typical two-stage stochastic linear program with fixed recourse as presented in section 3.1 in the book by Birge and Louveaux [15] and oftentimes applied in the economic system can be formulated as follows:

$$\min\{c^\top x + \mathbb{E}_\omega \left[\min q^\top(\omega)y(\omega) \,:\, Wy(\omega) = h(\omega) - T(\omega)x\,,\, y(\omega) \geq 0\,,\, \forall \omega \in \Omega \right],\, x \in \mathbf{X}\}$$

where $\mathbf{X}$ is a nonempty polyhedron (e.g. $\mathbf{X} = \{x \in \mathbb{R}^n \,:\, Ax = b, x \geq 0\}$), as well as $x \in \mathbb{R}^{n_1}$ and $y(\omega) \in \mathbb{R}^{n_2}$ for all $\omega \in \Omega$ are the decision vectors of the first and second stage, respectively. Let $(\Omega, \mathcal{A}, \mathcal{P})$ a probability space where Ω is a discrete or continuous set of outcomes, $\mathcal{A}$ a σ-algebra, and $\mathcal{P}$ a probability measure. The remaining vectors and matrices are data of appropriate sizes. If Ω is a continuous set, then the above

1. INTRODUCTION

equality $Wy(\omega) = h(\omega) - T(\omega)x$ and the inequality $y(\omega) \geq 0$ should hold $\mathcal{P}$-almost surely.

In order to get a proper two-stage scheme, nonanticipativity is assumend, i.e., first, x is chosen, then the outcome of the random vector

$$\xi(\omega) = (q^\top(\omega), h^\top(\omega), T_1(\omega), \ldots, T_k(\omega))$$

is observed where T_i are the rows of the matrix T, and not until then $y(x, \omega)$ is chosen dependent on the preceding values. Moreover, the distribution $\xi(\omega)$ must not depend on x. Instead of the expectation $\mathbb{E}$ another risk measure could be picked.

Let $\mathbf{Y}(x, \omega) = \arg\min_z \{q^\top(\omega) z : Wz = h(\omega) - T(\omega)x, z \geq 0\}$, then, the above model can be rewritten as

$$\min\{c^\top x + \mathbb{E}_\omega \left[q^\top(\omega) y(x, \omega)\right] : y(x, \omega) \in \mathbf{Y}(x, \omega), \forall \omega \in \Omega\right], x \in \mathbf{X}\}.$$

Compared to the linear stochastic bilevel program

$$\min\{c^\top x + \mathbb{E}_\omega \left[d^\top(\omega) y(x, \omega)\right] : y(x, \omega) \in \mathbf{Y}(x, \omega), \forall \omega \in \Omega\right], x \in \mathbf{X}\}.$$

the difference seems to be marginal since only the upper level cost vector for the lower level variable $y(x, \omega)$ has changed. But, this small difference has strong impacts on the complexity and the convexity of the problem. Besides, the question of the uniqueness of the solutions for the lower level becomes an issue. All this will be shown in the next section.

1.2 Basic Complexities

As could be seen in the last section, linear stochastic two-stage programs form a subclass of linear stochastic bilevel problems and, in addition, the complexity of bilevel problems rises compared to two-stage problems. For that purpose, it is assumend that Ω is discrete and finite, that is $|\Omega| = N$. Now that the random parameters are specified in the form of scenarios, the model of the linear two-stage problem is called the deterministic equivalent problem and can be displayed as

$$\begin{aligned} \min \quad & c^\top x + \sum_{k=1}^{N} \pi_k d_k^\top y_k \\ s.t. \quad & Ax = b, \\ & Wy_k = h_k - T_k x, y_k \geq 0 \quad k = 1, \ldots, N \end{aligned} \quad (1.2)$$

1.2 Basic Complexities

with $0 \leq \pi_k \leq 1$ being the weights of the different scenarios and $\sum \pi_k = 1$. This linear, convex programming problem is polynomially solvable since the input size of the model is just the size of all parameters (see e.g. Dyer and Stougie [38]).

On the other hand, the deterministic equivalent problem of the linear stochastic bilevel problem would be a linear bilevel model:

$$
\begin{aligned}
\min \quad & c^\top x + \sum_{k=1}^{N} \pi_k d_k^\top y_k(x) \\
s.t. \quad & Ax = b, \\
& y_k(x) \in Y_k(x), \quad k = 1, \ldots, N,
\end{aligned}
\tag{1.3}
$$

with $Y_k(x) = \arg\min_z \{q_k^\top z : Wz = h_k - T_k x, z \geq 0\}$.

Hansen et al. [48] showed in their paper that the linear bilevel problem is a strongly NP-hard problem. Actually, this property was even tightend by Vicente et al. [88]. They proved (based on a reduction from 3-SAT) that checking strict local optimality or just local optimality is already NP-hard. That means that even in the deterministic equivalent case, most likely, there will not exist polymial-time algorithms unless P = NP.

This subordination of the two problems can also be confirmed when analyzing the convexity of the problems. On the one hand, problem (1.2) is convex whereas problem (1.3) is normally neither convex nor differentiable (for more details see chapter 3).

In addition, let $q_1 = \cdots = q_N = q$ and

$$Q_k(x) = \min_y \{q^\top y : Wy = h_k - T_k x, y \geq 0\}$$

the recourse cost function of the deterministic equivalent of the linear two-stage problem with a fixed cost vector. Two assumptions will be made:

(A1) *[Complete Fixed Recourse]* The recourse matrix $W \in \mathbb{R}^{n_2 \times m_2}$ satisfies

$$\{z \in \mathbb{R}^{m_2} : z = Wy, y \geq 0\} = \mathbb{R}^{m_2}.$$

(A2) *[Sufficiently Expensive Recourse]* (Also called *Dual Feasibility*)

$$\{u \in R^{n_2} : W^\top u \leq q\} \neq \emptyset.$$

1. INTRODUCTION

It can be shown (see e.g. Birge and Louveaux [15]) that under these two assumptions the functions $Q_k, k = 1, \ldots, N$, are piecewise linear and convex in x. Thus,

$$Q_{\mathbb{E}}(x) := c^\top x + \sum_{k=1}^{N} \pi_k Q_k(x)$$

is also piecewise linear and convex in x. Here, $0 \leq \pi_k \leq 1$ represent the weights of each scenario and $\sum_{k=1}^{N} \pi_k = 1$.

But, under the same assumptions, the bilevel cost function

$$F(x,y) = c^\top x + \sum_{k=1}^{N} \pi_k d_k^\top y_k(x)$$

with $y_k(x) \in Y_k(x) = \arg\min_z \{q^\top z : Wz = h_k - T_k x, z \geq 0\}$, $k = 1, \ldots, N$, may be nonconvex in x. Just assume $d_k = -q$ for all $k = 1, \ldots, N$ and the bilevel problem reduces to a max-min-problem.

In contrast to stochastic programming, the uniqueness of the lower level solution set becomes a crucial issue here. Especially in the linear case, it can't be taken for granted that

$$Y_k(x) = \arg\min_y \{q^\top y : Wy = h_k - T_k x, y \geq 0\}$$

is a singleton. So, the leader has to decide about his risk attitude in conjunction with his relation to the follower. If the relation to the follower is friendly or cooperative (e.g. in some prinicipal-agent settings or when optimizing decision questions arising in a decentralized company), the **optimistic** approach would be recommendable. It is assumed that in case of multiple solutions, the follower chooses that variable that is best for the leader, i.e., the following vector is chosen for every $k = 1, \ldots, N$

$$y_k^o(x) = \arg\min_z \{\pi_k d_k^\top z : z \in Y_k(x)\}.$$

On the other hand, if the situation is rather competitive, the **pessimistic** solution concept would be a good alternative. For the follower's reaction, the worst case is assumed, that is

$$y_k^p(x) = \arg\max_z \{\pi_k d_k^\top z : z \in Y_k(x)\}.$$

1.3 Motivation

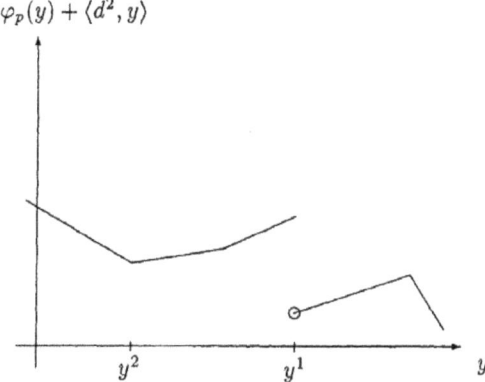

Figure 1.1: Possible form of a pessimistic bilevel value function, found in Dempe [28].

Both approaches have been developed early in literature but the optimistic is mostly applied in economics. That is because the pessimistic bilevel problem may not have an optimal solution due to the fact that the pessimistic upper level value function

$$F^p(x) = c^\top x + \sum_{k=1}^{N} \pi_k d_k^\top y_k^p(x)$$

may not be lower semicontinuous. The absence of this property includes that the pessimistic bilevel problem may not have an optimal solution. A deterministic example for this can be seen in figure 1.1 where y represents the upper level variable in this case and $\varphi(y) + d^{2\top} y = F^p(y)$.

Besides, most reformulations of bilevel problems – in order to compute solutions – rely on the optimistic case. So, throughout this thesis, the optimistic approach will be used.

1.3 Motivation

In spite of the differences, two-stage and bilevel stochastic problems bear a lot of structural similarities. Still, only few papers were published that make use of decomposition methods, see e.g. Lin et al. [61] where a Dantzig-Wolfe decomposition method is used

1. INTRODUCTION

in a heuristic to solve bilevel dynamic network design problems. But, the structural resemblance suggests the possibility that those techniques can particularly be used in order to find global optimal solutions for linear stochastic bilevel problems. More informations about methods known from stochastic programming will be given in chapter 2.

Due to the growing performance of computers, stochastic bilevel problems become more and more solvable and, thus, interesting for economical benefit. Amongst others, utility companies have supported some practical research in exchange for optimal offering or pricing strategies. Ruiz and Conejo adress in their paper [78] the problem of finding the optimal offering strategy for a price-making power producer that trades electric energy in a day-ahead electricity pool. Besides, uncertainty associated with demand bids and generating offers – both appearing scenariowise in the second level – is considered. The problem is linear except for the leader's objective where a product of leader and follower variables occur. This term is linearized and the second level is firstly replaced by its Karush-Kuhn-Tucker conditions. Then, the newly built complementarity conditions are remodelled using binary variables and the big-M method. The resulting mixed-integer linear one-level problem is then solved using ILOG CPLEX [53]. As the authors ascertain themselves, the problems of this procedure lies within the fact that

- the computational time increases dramatically if scenarios are considered (in their case study, the computational time of eight scenarios compared to the deterministic case took about 30 times longer)

- the appropriate selection of the big-M parameters generally requires a "nontrivial trial-and-error process" and, thus, can also increase the computational time.

The algorithm presented in chapter 6 is in fact based on the reformulation of the big-M method but does not depend on the calculation of the parameters.

1.4 Literature

Despite their closeness to Stackelberg games which have been known for quite some time, bilevel programming problems have gained mathematical interest in the early 1980's, only.

1.4 Literature

In rigorous mathematical terms, the problem was introduced by Bracken and McGill in 1973, see [16]. However, Candler and Norton [18] were the first to name the problems *bilevel* and *multilevel*. Still, there are more recent works from authors that name their problems two-level instead of bilevel. Nowadays, the term bilevel is established.

Surveys

In the early 80's, a lot of effort has been spent by the mathematical community in order to analyze different kinds of bilevel and multilevel problems. A good overview is given in 1994 by Vicente and Calamai [89]. Linear bilevel problems were reviewed by Wen and Hsu [91] in the year 1991.

Another survey with numerous references is given in 2003 by Dempe [29]. It includes a short overview of bilevel programming and mathematical problems with equilibrium constraints.

The most recent overview of bilevel programming was presented by Colson et al. [26] in 2007. Beside solution approaches and applications, it focuses on the connection between bilevel models and mathematical problems with equilibrium constraints.

Monographs

Two books on bilevel programming were published so far: the first was written by Bard [7] and the next by Dempe [28]. Bard concentrated his research on different algorithms to solve bilevel problems as well as bilevel applications. Dempe's book, on the other hand, is very extensive due to the fact that it is concerned with applications, linear, discrete and nonlinear bilevel problems, parametric optimization as well as optimality conditions and different solution algorithms. Many of the important results for linear and other bilevel programs can be found in the latter. It is so far the most recent book in that research area but does not include stochastic bilevel problems. The remainder of this section will thus focus only on stochastical bilevel publications.

Stochastic Bilevel Articles

The problems considered below are all optimistic if not mentioned otherwise.

Patriksson and Wynter [71] were the first to publish a paper concerning stochastic bilevel problems in 1999. Therein, bilevel, equilibrium constrained, and two-stage problems with recourse are presented along with their discretely and continuously distributed stochastic versions. Some relations between these models are presented and – more importantly – conditions for the existence of solutions, convexity and directional differentiability are derived. The paper concludes with the presentation of different

1. INTRODUCTION

ideas how to solve stochastic mathematical programs with equilibrium constraints algorithmically, including a descent and a penalty algorithm as well as suggestions for scenario decomposition. The equilibrium constraint in this paper is described by a variational inequality, but the bilevel problem is presented in its traditional way. In all models, the lower level or second stage variable just infects the upper level objective function, but not its constraints. Additionally, there are no approaches made for the case that the lower level solution is not unique. A forerunner to this report was the conference presentation "Bilevel stochastic programming for network equilibrium problems" at the International Symposium on Mathematical Programming (ISMP) in 1997 by Wynter.

Both, Wynter and Patriksson kept a research focus on stochastic bilevel programming and its application in network design problems as well as structural optimization. Patriksson himself is giving an overview on papers concerning stochastic mathematical programs with equilibrium constraints (SMPEC) published between 1997 and 2008 in section 1.3 of his paper [70]. The rest of this article is concerned with continuously distributed traffic SMPECs and the study of the dependence of optimal solutions on the probability distribution. Therefore, the author also distinguishes between the cases where the upper level constraints are joint, that is, simultaneous constraints in both variables, x and y, and where these constraints are only in x. It is shown there that under certain strong assumptions and in the absence of joint upper level constraints the probability distribution are continuous in the optimal design x (upper level variables).

At the end of his work, Patriksson presents two inexact exterior penalty methods, one regarding the upper level, the other one regarding the lower level constraints. Besides, two discretizations of the probability space are presented, one based on the method of mechanical quadratures and the other is an application of sample average approximation.

The remainder of this section is concerned with publications in the field of stochastic bilevel problems that are not mentioned in Patriksson [70]. Of course, it gives only an insight.

The doctor's thesis [92] by Werner, published in 2005, supervised by Gaivoronski, was one of the first to deal with stochastic bilevel problems. The thesis mainly consists of four papers, of which one is published (see Audestad, Gaivoronski, and Werner [4]). The first paper [4] is concerned with building a stochastic two-stage bilevel model

1.4 Literature

with a finite number of scenarios for a network operator in the telecommunication environment. The two stages are realized in the lower level. The leader chooses his periodic decisions and the follower can just react to the stochastic outcomes and the decision of the leader. The problems are linear-constrained problems with quadratic, concave objective functions. A deterministic equivalent formulation is given.

The second paper introduces the uncertainty more thoroughly, necessary optimality conditions of Fritz-John type are given and an algorithm is presented utilizing a stochastic quasi-gradient method. At first, the uncertainty is modelled as a two-stage problem in the upper level which also exhibits constraints in the follower's variables. Thus, the feasibility region can be disconnected. Assumptions are made such that the leader's and follower's objective functions are convex and differentiable and the follower's optimal solution is always uniquely determined. An one-level two-stage reformulation is given using the activity sets of the lower level constraints and KKT conditions. This problem is solved by a stochastic quasi-gradient method which works on different segments of the feasibility region. It is shown that, under certain assumptions, the algorithm converges.

In a second approach, not only the leader, but also the follower can incorporate a second stage decision. However, "the follower does not regard the future when she determines her action [..] in the first stage, i.e., her second stage problem is not interpreted as recourse problem" (see Werner [92], page 84). Again, under certain assumptions the problem can be reformulated as the above descripted one-level two-stage problem and solved with the same algorithm. Numerical studies are given (for two different step sizes) where the algorithm needs more than 2000 iterations for a small stochastic problem. In all models, Werner uses the expectation as risk functional.

The third paper is concerned with the utilization of stochastic bilevel programming and agency problems in economy; and the fourth paper presents conditions under which the stochastically perturbed function $F(z) = \mathbb{E}_\omega[f(z+\omega)]$ becomes strictly convex while f was only convex. Thus, an alternative to the pessimistic and optimistic approaches is given that allows for unique lower level solutions when including stochasticity. A follow-up paper of Gaivoronski and Werner [45] in the framework of agency problems also deals with modeling questions and solution approaches.

Roghanian et al. [77] consider a bilevel and multiobjective problem with joint chance constraints for supply chain planning. The joint chance constraints are instantly re-

1. INTRODUCTION

placed by their deterministic equivalent using Hulsurkar et al. [52]. Several numerical examples are solved using the algorithm for suchlike deterministic problems proposed by Osman et al. [68].

Carrión et al. [20] present a (stochastic bilevel) decision-making framework for an electricity retailer along with computational behavior of the resulting mixed-integer linear model using a commercial solver. Both levels consist of two-stage problems resulting in here-and-now decisions for the leader and the follower.

A comparable approach for the gas market has been made by Kalashnikov et al. [56] and Dempe et al. [34]. Whereas, the former paper presents a mixed-integer problem, numerical results, as well as a comparison with the perfect information solutions and the expected value solutions (in order to evaluate the benefit of the introduction of stochastics). The latter uses an inexact penalization approach in order to formulate a new algorithm which is shown to converge. The latter algorithm is evaluated and compared to existing methods.

Fampa et al. [40] present a stochastic bilevel problem for bidding at electricity markets. The paper also includes a primal-dual heuristic which is evaluated and compared to other solvers.

A linear mixed-integer bilevel problem with a probabilistic knapsack constraint is analyzed by Kosuch et al. [59]. The finite sample space allows to reformulate the problem deterministically. Two reformulations follow and the authors are left with a linear deterministic single-level problem with complementarity constraints. The remaining problem is then solved using Lagrangian relaxation for the quadratic terms and an iterative scheme using upper and lower bounds for the Lagrangian term. Numerical results are given where the computing time is not dependent on the number of scenarios.

The doctoral thesis by Pisciella [74] discusses different business models for service providers, two of them being stochastic bilevel problems where the uncertainty lies in the future demand in the lower level. For both models, two approaches are made. The first approach uses finitely many scenarios and results in a linear mixed-integer program, whereas the second works with a continuous distribution which causes the optimal value function of the lower level to be differentiable under weak assumptions. The expected value is chosen to evaluate the uncertainty. Both approaches are solved and numerical results are evaluated.

1.4 Literature

A paper by Alizadeh et al. [2] deals with a bilevel problem over a transportation network where both levels are two-stage problems. A discrete distribution is assumed and the problem is reformulated as a single-level problem. Some properties are analyzed along with some examples.

Very recently, Ivanov [55] focused on linear stochastic bilevel problems whose objective is given by a Value-at-Risk (see chapter 2). The uncertainty appears in the right-hand side of the follower's problem. Sufficient conditions are given such that the risk funcitonal is a Lipschitz-continuous function of the upper level decision. The problem is reformulated into an one-level two-stage problem using the KKT conditions. Numerical experiments are reported for discretely distributed problems, with up to 25 realizations and decision variables of dimensions two in the upper and three in the lower level.

Kovacevic and Pflug [60] are concerned with electricity swing option pricing in the monopolistic and the competitive case. Both cases are modelled as bilinear bilevel problems with multistage decisions in the lower level while the expectation is used for the first case. The second uses the Conditional Value-at-Risk. Simple algorithms are proposed that make use of the special structure, including one algorithm that is capable to solve a pessimistic instance. Numerical examples are given along with a concise overview on related algorithms.

Interestingly, most of the previous mentioned papers were motivated by network-related problems, arising in telecommunications, electricity markets or transportations. These problems inherit a natural order of successive decision making under uncertainty and constitute, at the moment, the biggest field of applications for stochastic bilevel problems.

In the case when the lower level can be replaced by its Karush-Kuhn-Tucker conditions (see chapter 5), the bilevel problem can be reformulated as a mathematical program with complementarity contraints (MPCC). Therefore, also papers on stochastic MPCCs are of importance here. A state of the art survey of 2009 is given by Lin and Fukushima [62] containing a few references on papers for stochastic MPCCs.

A paper by Birbil et al. [14] discusses a sample-path method (also known as sample average approximation) for stochastic MPCCs where some functions are replaced by the expectation of these functions supplementary dependent on a random event ω. All decisions are here-and-now decisions since they have to be made before the random

1. INTRODUCTION

event is realized. Also, other papers by two of the authors – Birbil and Gürkan – were published in that field but are already mentioned in Patriksson [70].

A stochastic MPCC for the electricity market is given in Zhang et al. [99]. The authors examine existence and uniqueness of a Nash-Cournot equilibrium, which is realized as a complementarity constraint, along with other properties of the model.

Liu et al. [63] as well as Xu and Ye [93] consider a two-stage problem where the second stage comprises a complementarity constraint along with other constraints. Some constraint qualifications are examined as well as the stability of the problem in order to be approximated by "ordinary" two-stage stochastic nonlinear programs. In the former paper, a sample average approximation is conducted, whereas the latter focuses on optimality conditions.

None of the presented algorithms for stochastic bilevel problems or stochastic MPCCs made use of a scenariowise decomposition technique.

2
Linear Two-Stage Problems

As already stated in the work of Patriksson and Wynter [71] and verified through the reformulation in section 1.1, linear two-stage problems are a special case of linear stochastic bilevel problems. Still, the former gained considerably more research interest than the latter. This chapter will give a very brief discussion. More information can be found, e.g., in the monographs Birge and Louveaux [15], Censor and Zenios [22], Kall and Wallace [57], or Prékopa [75]. Also, a broad collection of publications in that field can be found on the internet, collected by van der Vlerk [86].

Many disciplines, such as economics, mathematics, and statistics, have put time and effort in analyzing stochastic programs. Therefore, the applications vary from financial planning over agricultural questions to logistic problems and more. Every time data is not known for certain but can be approximated by a discrete or continuous probability distribution, the user can profit from developments in that field. Numerous approaches and solution methods were developed among which probabilistic constraints, two-stage problems, and decomposition methods inherit an important role.

Probabilistic constraints, also known as chance constraints, allow the decision maker to determine to what percentage at least (or at most) a constraint has to hold, dependent on some probability measure. An example is

$$\mathcal{P}(\{\omega \in \Omega : g(x,\omega) = 0\}) \geq \alpha$$

where $\mathcal{P}$ is a probability measure, $\omega \in \Omega$ a random event, $g : X \times \Omega \to \mathbb{R}$ a proper function, X some Banach space, and $0 < \alpha \leq 1$ a real value that is called probability level. These individual chance constraints can be part of different kinds of optimization

2. LINEAR TWO-STAGE PROBLEMS

models where x might be a discrete or continuous variable. If possible, the inverse of the distribution function or the variance is used in order to reformulate the constraint into computationally or mathematically simpler ones (see Birge and Louveaux [15]).

However, if the same decision vector x occurs in a system of random constraints, they are called joint probabilistic constraints. These constraints are much harder to treat and feasible solutions x are difficult to find because the resulting problems might be nonconvex and the feasibility domain might not even be connected dependent on the distribution function. In order to solve suchlike problems, often, piecewise approximations of the inverse of the distribution function are executed.

Another possibility to incorporate uncertainty is via a recourse program in which a first decision is made before the uncertainty is disclosed and some recourse action can be taken afterwards. This line of action is also called *nonanticipativity*. It forces the first decision variable x to be taken without the anticipation of future events ω. On the other hand, the recourse action y is dependent on x as well as on ω. Its function is to compensate the random event.

A linear recourse problem was already presented in section 1.1. This representation was chosen in order to make the comparison with bilevel problems easier. The same model can be represented using

$$\phi(t_1, t_2) = \min\{t_1^\top y \ : \ Wy = t_2, \, y \geq 0\} \qquad (2.1)$$

the value function of the inner linear program and

$$f(x, \omega) = c^\top x + \phi(q(\omega), h(\omega) - T(\omega)x)$$

the random total cost function. Then, the problem can be stated as

$$\min\{\mathbb{E}_\omega(f(x, \omega)) \ : \ x \in \mathbf{X}\}.$$

These definitions also allow for different functions than affine or linear ones, e.g. bilinear terms, quadratic random cost, or nonlinear constraints. If ϕ is unbounded below or infeasible, the value of the second stage problem is defined to be $-\infty$ or $+\infty$, respectively. Additionally, if Ω is a finite set, the expected value is then the weighted sum of the functions $\phi_k(x) = \phi(q_k, h_k - T_k x), k = 1, \ldots, N = |\Omega|$. For that case, it is $+\infty - \infty = +\infty$ in order to reject any first stage variable that produces an infeasible solution.

The problem (2.1) has fixed recourse, i.e., the recourse matrix W is not random. Elsewise, only few can be said about solutions and their existence (see Gollmer [46] and Freund [43]).

Taking the expectation in the above model may lead to solutions x^* whose random variable $f(x^*, \omega)$ takes big values more frequently than wanted by the user. Instead, other risk functionals can be exerted such as the excess probability (see e.g. Bereanu [10]), the Value-at-Risk, or the Conditional Value-at-Risk (for both see e.g. Pflug [72], or for the latter see Rockafellar and Uryasev [76]). These functionals or measures reflect a rather risk-averse attitude given that they evaluate worst outcomes.

In these cases, the optimization is based on a weighted sum of the expectation and some other risk measure $\mathcal{R}$, i.e.,

$$\min\{\underbrace{\mathbb{E}_\omega(f(x,\omega)) + \rho \cdot \mathcal{R}(x)}_{=Q_{MR}(x)} : x \in \mathbf{X}\}$$

where $\rho \geq 0$ is some fixed parameter. The above mentioned risk functionals are:

- **Excess Probability:**

$$\mathcal{R}(x) = Q_\eta(x) := \mathcal{P}(\{\omega \in \Omega : f(x,\omega) > \eta\})$$

This risk functional describes the probability of exceeding a prescribed threshold η.

- **Value-at-Risk:**

$$\mathcal{R}(x) = Q_{\alpha VaR}(x) := \inf\{\eta : \mathcal{P}(\{\omega \in \Omega : f(x,\omega) \leq \eta\}) \geq \alpha\}$$

It expresses for a given probability level $0 < \alpha \leq 1$ the best (i.e., smallest) outcome of the $(1-\alpha) \cdot 100\%$ worst. It is also called the α-quantile and can be presented as the inverse $G^{-1}(\alpha)$ of the distribution function $G(u) = \mathcal{P}(\{\omega \in \Omega : f(x,\omega) \leq u\})$.

- **Conditional Value-at-Risk:**

$$\mathcal{R}(x) = Q_\alpha(x,\eta) := \frac{1}{\alpha}\int_0^\alpha G^{-1}(u)du$$

The Conditional Value-at-Risk (CVaR) is the expected value of the costs in the $\alpha \cdot 100\%$ worst cases, for a given probability level $0 < \alpha < 1$.

2. LINEAR TWO-STAGE PROBLEMS

It is also called Average Value-at-Risk or Expected Shortfall because it occurs as the average of the Value-at-Risks Q_{uVaR} for $0 \leq u \leq \alpha$.

Schultz [81] showed that the excess probability as well as the Conditional Value-at-Risk possess a binary-linear and linear equivalent formulation, respectively, if a discrete distribution can be assumed and the uncertainty lies only in the vector on the right-hand side.

More information about these and other risk functionals can be found in the book by Pflug and Römisch [73].

2.1 Properties

Unlike in bilevel programming, only the optimal value of the second stage is of interest in stochastic programming. That gives rise to further study the function $\phi : \mathbb{R}^m \times \mathbb{R}^r \to \mathbb{R}$.

Theorem 2.1 (Nožička et al. [67])
Let
$$K_1 := \left\{ t_1 \in \mathbb{R}^m : \{u \in \mathbb{R}^r : W^\top u \leq t_1\} \neq \emptyset \right\}$$
and
$$K_2 := \left\{ t_2 \in \mathbb{R}^r : \{y \in \mathbb{R}^m_+ : Wy = t_2\} \neq \emptyset \right\}.$$
Then, the following holds:

a) $\phi(t_1, t)$ is piecewise linear, continuous and concave in $t_1 \in K_1$ for all fixed $t \in K_2$.

b) $\phi(t, t_2)$ is piecewise linear, continuous and convex in $t_2 \in K_2$ for all fixed $t \in K_1$.

c) $\phi(t_1, t_2)$ is continuous on $K = K_1 \times K_2$ and there exists a finite partition of K into s cones $K^j, j = 1, \ldots, s$, each of dimension $m+r$ such that $\phi(t_1, t_2)$ is bilinear on K^j.

So, under the assumptions of complete fixed recourse and sufficiently expensive recourse (see page 7 in the previous chapter), the function $\phi(q, h - Tx)$ is piecewise linear and convex in x for any q, h, and T and so is $f(x) = c^\top x + \sum_k \phi(q_k, h_k - T_k x)$.

2.2 Decomposition Methods

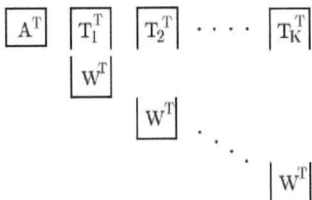

Figure 2.1: Dual block-angular structure, see Birge and Louveaux [15].

2.2 Decomposition Methods

There are two big classes of solution methods that have been developed for high dimensional linear problems with a special structure. The first one was developed by Dantzig and Wolfe [27] and can be applied to problems with a block structure as in figure 2.1. It uses a technique of delayed column generation as the problem is split into a master problem and one or more subproblems. Columns (decision variables) in the master program are added or replaced during the procedure.

The other one was developed by Benders [9] two years later and can be applied to the dual form, figure 2.2. This procedure uses a form of row generation. It was applied and specialized to stochastic problems by Van Slyke and Wets [87].

Dantzig-Wolfe-Decomposition

For illustration, assume that the problem to be solved has the form

$$\min_x c^\top x \quad \text{subject to} \quad Ax = b, \quad l \leq x \leq u$$

with A being a $m \times n$ matrix. The matrix A can then be partitioned into two matrices A' and A'' of dimension $m' \times n$ and $m'' \times n$, respectively. The above problem is then equivalent to

$$\min_x c^\top x \quad \text{subject to} \quad A'x = b', \quad A''x = b'', \quad l \leq x \leq u.$$

For the relaxed feasibility region

$$P = \{x \in \mathbb{R}^n : A''x = b'', \quad l \leq x \leq u\}$$

2. LINEAR TWO-STAGE PROBLEMS

there exists a finite number of basic feasible solutions or extreme points $v_1, \ldots, v_M \in \mathbb{R}^n$ as well as a finite number of basic feasible directions (extreme rays) $w_1, \ldots, w_N \in \mathbb{R}^n$ such that every $x \in P$ can be described as

$$x = \sum_{k=1}^{M} r_k v_k + \sum_{j=1}^{N} s_j w_j$$

with $r_k, s_j \geq 0$ and $\sum r_k = 1$. Now, x can be substituted in the above problem:

$$\begin{aligned}
\min_{r_k, s_j} \quad & \sum_{k=1}^{M} r_k c^\top v_k + \sum_{j=1}^{N} s_j c^\top w_j \\
s.t. \quad & \sum_{k=1}^{M} r_k A' v_k + \sum_{j=1}^{N} s_j A' w_j = b', \\
& \sum_{k=1}^{M} r_k = 1, \\
& r_k \geq 0, \quad s_j \geq 0
\end{aligned} \quad (2.2)$$

This is the so-called (complete) master problem. In the algorithm, not all extreme points and rays will be put into the reduced master problem, thus, there are also less decision variables. Only those basic feasible solutions and directions are of interest that produce a better objective value. These vectors can be found using the pricing problem

$$\min_x \ (c - A'^\top y)^\top x \quad \text{subject to} \quad A''x = b'', \ l \leq x \leq u$$

where y is the dual solution of (the reduced) master problem (2.2).

If the problem has the block angular structure as in figure 2.1, A' corresponds to the first block of constraints $(A^\top, T_1^\top, \ldots, T_K^\top)$ and the algorithm will start with finding a feasible basic solution to $A'x = b'$ and its dual solution y – in order to then solve the K subproblems. The master program incorporates one or all of the new columns generated by the solutions of the subproblems (based on their ability to improve the original problem's objective). If the objective of the master program is improved, the algorithm starts again with finding the dual solution to this program and solving the subproblems. Otherwise, the algorithm stops since the master program cannot be improved by any solution of the subproblems.

The notation was taken from Chvátal [24]. Further information can be found in Bertsimas and Tsitsiklis [11].

L-shaped Decomposition

Generally, this procedure – based on the work by Van Slyke and Wets [87] – can be applied to the deterministic equivalent model (1.2) since this has the so-called L-shape.

2.2 Decomposition Methods

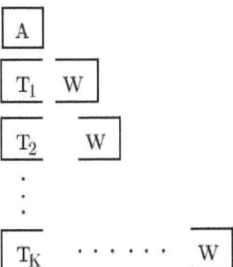

Figure 2.2: Block-angular structure, also called L-shaped, see Birge and Louveaux [15].

The idea behind this cutting plane algorithm is to partition the set of decision variables in two sets of first stage variable x and second stage variables $y_i, i = 1, \ldots, K$, only solve a problem for the first stage variables (and a reduced number of constraints, a simpler problem), include the optimal value $\hat{x}$ to subproblems each concerning one y_i and adding new constraints with information from these subproblems to the simpler problem.

In particular, the algorithm will start with the reduced problem

$$\min_x \quad c^\top x \quad \text{subject to} \quad Ax = b, x \geq 0 \qquad (2.3)$$

which will be extended during the procedure with only one variable θ in order to represent the objective value of the second stage variables as well as new constraints in x and θ in order to cut off infeasible or suboptimal solutions.

The K subproblems that will produce the optimality cuts are of the form

$$\begin{aligned}\min_y \quad & w = q_k^\top y \\ \text{s.t.} \quad & Wy = h_k - T_k \hat{x} \\ & y \geq 0\end{aligned} \qquad (2.4)$$

where T_k, h_k, and $q_k, k = 1, \ldots, K$ are the different values for each scenario. If complete or relatively complete recourse ($h_k - T_k x \in posW := \{t : Wy = t, y \geq 0\}$ for all k and x) cannot be assumed, it might happen that one or more of these problems are not solvable. In that case, a program similar to that of phase 1 of the simplex method can

2. LINEAR TWO-STAGE PROBLEMS

be used in order to construct feasibility cuts:

$$\begin{aligned}\min_{y,v^+,v^-} \quad & w' = e^\top v^+ + e^\top v^- \\ \text{s.t.} \quad & Wy + Iv^+ - Iv^- = h_k - T_k\hat{x} \\ & y, v^+, v^- \geq 0,\end{aligned} \quad (2.5)$$

where e is a row vector of 1's and I is the identity matrix of appropriate size. This problem has always an optimal solution with $w' \geq 0$. If $w = 0$, the above problem (2.4) has at least one feasible solution, but if $w' > 0$ a feasibility cut has to be induced on x in order to cut off those x that produce infeasible second stages.

For both problems, the simplex multipliers (or dual solutions) are used in order to get cuts only in x. For the feasibility cut, the optimal dual solution σ of problem (2.5) fulfills

$$\begin{aligned}\min w' &= \sigma[h_k - T_k\hat{x}] \\ \sigma W &\leq 0 \\ -e \leq \sigma &\leq e.\end{aligned}$$

If $w' = \sigma[h_k - T_k\hat{x}] > 0$, then $\{t : \sigma t = 0\}$ is a hyperplane that separates $h_k - T_k\hat{x}$ and $posW$. In order to produce feasible second stages, both must be on the same side, thus $\sigma[h_k - T_k\hat{x}] \leq 0$ has to hold. This constraint can be added to problem (2.3).

Now, to explain the optimality cuts, I will rewrite problem (1.2) using the value function of the inner problem

$$\begin{aligned}\min_{x,\theta} \quad & c^\top x + \theta \\ \text{s.t.} \quad & Ax = b \\ & \theta \geq \sum_{k=1}^{K} p_k \phi(q_k, h_k - T_k x) \\ & x \geq 0.\end{aligned} \quad (2.6)$$

If all problems (2.4) are feasible for $x = \hat{x}$, then the dual optimal solution π_k will fulfill

$$\pi_k[h_k - T_k\hat{x}] = \phi(q_k, h_k - T_k\hat{x})$$

and, thus, $\pi_k[h_k - T_k x]$ is a support of $\phi(q_k, h_k - T_k x)$ (a complete proof can be found in Van Slyke and Wets [87]). Consequently, a feasible tuple (x, θ) must meet

$$\theta \geq \sum_{k=1}^{K} p_k(\pi_k h_k - \pi_k T_k x) = \sum_{k=1}^{K}(p_k \pi_k h_k) - (p_k \pi_k T_k)x. \quad (2.7)$$

The complete algorithm can be found in Van Slyke and Wets [87] or Birge and Louveaux [15].

3
Linear Stochastic Bilevel Problems

Linear bilevel problems bear a lot of difficulty already in their deterministic form. That is why the deterministic program and its properties will be presented primarily. Stochastics will first be introduced in section 3.2 on page 40.

If the set Ω of random events is finite ($|\Omega| = N$), the lower level multifunction $Y(x, \omega)$ can be indexed by the random events, with $i = 1, \ldots, N$, and for every such random event an optimal solution $y_i(x)$ can be chosen from $Y_i(x)$ to optimize the leader's objective. In that case, the constraint set of the bilevel problem consists of multiple second levels. Defining $y(x) = (y_1(x), \ldots, y_N(x))$ and $Y(x) = Y_1(x) \times \ldots \times Y_N(x)$ shows, then, that all properties to be shown for the deterministic linear bilevel problems also hold for stochastic linear bilevel problems with finite random sets.

For finite random sets in two-stage programming, similar constructions apply. The remaining problem is called *deterministic equivalent*.

3.1 Linear Bilevel Problems

For readability, the considered program is displayed again:

$$\begin{aligned} \min \quad & c^\top x + d^\top y \\ \text{s.t.} \quad & Ax = b \\ & y \in Y(x) = \arg\min_{y^*}\{q^\top y^* : Wy^* + Hx = h, y^* \geq 0\}. \end{aligned} \quad (3.1)$$

3. LINEAR STOCHASTIC BILEVEL PROBLEMS

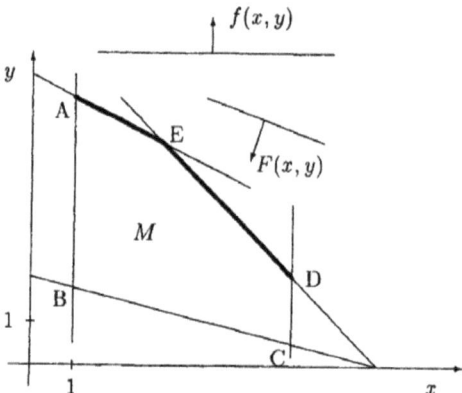

Figure 3.1: The feasible set of the bilevel problem is only described by the two thick-lined edges connecting A and E, E and D, respectively, found in Dempe [28].

In order to show the lack of convexity of this kind of problem, an example as shown in Dempe [28] will be employed.

Example 3.1 (Dempe [28])
The decision vectors are each one-dimensional and the problem is to

$$\begin{aligned}
\min \quad & x + 3y = F(x,y) \\
\text{s.t.} \quad & 1 \leq x \leq 6 \\
& y \in Y(x) = \arg\min_{y^*}\{-y^* \,:\, x + y^* \leq 8,\ x + 4y^* \geq 8,\ x + 2y^* \leq 13\}
\end{aligned}$$

The optimal solution of the lower level can be described by means of a distinction of cases of the upper level variable:

$$y(x) = \begin{cases} 6,5 - 0,5x & \text{if } 1 \leq x \leq 3, \\ 8 - x & \text{if } 3 \leq x \leq 6 \end{cases}$$

In figure 3.1, M denotes the set of all pairs (x, y) satisfying the constraints of both levels and f is the lower level cost function $f(x, y) = -y$. Only the thick line connecting the points A, E and D represents the set of all feasible solutions of this bilevel problem. Thus, the global optimal solution is point $D = (6, 2)$ with objective function value $F(6, 2) = 12$.

3.1 Linear Bilevel Problems

The displayed example is a very simple one but shows already that the set of feasible solutions is in general nonconvex, but indicates that the latter might be connected. This is, in general, true as will be shown in the next theorem. But, if there would be constraints in the upper level that contain the followers variable y, then, the solution set might even be disconnected. In example 3.1, the additional upper level constraint

$$x + \tfrac{5}{4} y \leq 9$$

would cut off point $E = (3, 5)$ and a small neighborhood of it (since $3 + 6\tfrac{1}{4} = 9\tfrac{1}{4} > 9$), but not point $D = (6, 2)$ $(6 + 2\tfrac{1}{2} = 8\tfrac{1}{2} \leq 9)$ and also not $A = (1, 6)$ $(1 + 7\tfrac{1}{2} = 8\tfrac{1}{2} \leq 9)$. Thus, there would be a part in the lower level feasibility and optimality sets that produces infeasible solutions in the upper level for x around 3. It should be noted that the feasibility set of the problem would be connected again if the additional constraint would move into the lower level problem.

Theorem 3.2 (see Parraga [69])
The linear bilevel problem (3.1) is in general nonconvex, nondifferentiable, but connected for all those x for which the lower level is feasible. □

In the above example, the optimal solution set is unique, i.e.,

$$Y(x) = \arg\min_{y}\{q^\top y : Wy = h - Tx, y \geq 0\}$$

is a singleton for all x. In general, especially in higher dimensions, this is not the case, see for instance:

Example 3.3
The problem now is to
$$\max_{x} \quad x_1 + x_2 + 2y_1 + 3y_2$$
$$\text{s.t.} \quad 0 \leq x_i \leq 2, \, i = 1, 2$$
$$y \in Y(x) = \arg\max_{y^*}\{y_1^* + y_2^* : y_1^* \leq x_1, \, y_2^* \leq 2x_1, \, y_1^* + y_2^* \leq x_2, \, y_1^*, y_2^* \geq 0\}$$

The optimal solution set of the follower can then be characterized as

$$Y(x) = \begin{cases} y_1 = 0, \, y_2 = 0, & \text{if } x_1 = 0 \vee x_2 = 0, \\ y_1 = x_1, \, y_2 = 2x_1 & \text{if } x_2 \geq 3x_1 > 0, \\ \{y_1, y_2 \in \mathbb{R}_+ : y_2 = x_2 - y_1, \, y_1 \leq x_1, \, y_2 \leq 2x_1\} & \text{if } 3x_1 > x_2 > 0 \end{cases}$$

3. LINEAR STOCHASTIC BILEVEL PROBLEMS

In the latter case, it is not clear which solution can be chosen by the leader and the problem is not well-posed. In order to regain this property, the leader has to appraise the actual reaction of the follower. As already stated in section 1.2, there are two approaches developed in the literature for this case, the optimistic and the pessimistic.

In the optimistic case, the leader can assume a cooperative relationship between himself and the follower, and the problem would be to

$$\max_x \max_y \quad x_1 + x_2 + 2y_1 + 3y_2$$
$$\text{s.t.} \quad 0 \leq x_i \leq 2,\, i = 1, 2$$
$$y \in Y(x)$$

The overall optimal solution to this problem is $x = (2,2)^\top$ and $y_o = (0,2)^\top$ with the objective value 10.

In the pessimistic case, the problem is stated as

$$\max_x \min_y \quad x_1 + x_2 + 2y_1 + 3y_2$$
$$\text{s.t.} \quad 0 \leq x_i \leq 2,\, i = 1, 2$$
$$y \in Y(x)$$

For this problem, there exist multiple optimal solutions with objective value 8. In the optimal solution, it always holds $x_2 = 2$ and for $\frac{2}{3} \leq x_1 \leq 2$, the lower level pessimistic solution is $y_p = (x_1, 2 - x_1)^\top$.

This example showed that the pessimistic and optimistic approaches differ considerably.

The pessimistic leader's objective function $F^p(x)$ is in most cases not lower semicontinuous as can be seen in the figure 3.2.

In the linear case, when the leader's variable only influences the right-hand side of the follower's problem, the pessimistic optimal value function

$$F^p(x) = c^\top x + \sum_{k=1}^{N} p_k \max_z \{d_k^\top z \,:\, z \in Y_k(x)\}$$

can at least be shown to be upper semicontinuous. In order to show that and the lower semicontinuity for the optimistic case $F^o(x)$, a few definitions and theorems are needed that can be found in Bank et al. [5].

28

3.1 Linear Bilevel Problems

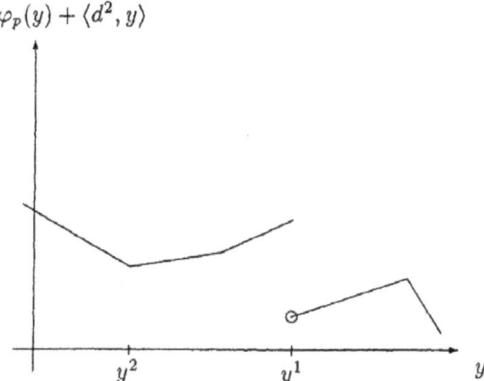

Figure 3.2: Possible form of a pessimistic bilevel value function, found in Dempe [28].

Definition 3.4
A function $f : \mathbb{R}^n \to \mathbb{R}$ is **weakly analytic** if for all $x, u \in \mathbb{R}^n$ it holds:

If the function $f_{x,u}(\alpha) = f(x + \alpha u)$ is constant on an open interval $[\underline{\alpha}, \overline{\alpha}] \subset \mathbb{R}$

$$\Rightarrow \quad f_{x,u}(\alpha) = f(x) \quad \forall \alpha \in \mathbb{R}$$

Obviously, all linear and affine functions are weakly analytic.

Definition 3.5
A point-to-set mapping (or multi-function) $\Gamma : \mathbb{R}^n \to 2^{\mathbb{R}^m}$ (where $2^{\mathbb{R}^m}$ is the power set of $\mathbb{R}^m$) is called

1. **closed** if for all sequences $(x_n) \subset \mathbb{R}^n$, $x_n \to x_0$, $(y_n) \in \Gamma(x_n)$, and $y_n \to y_0$, it holds $y_0 \in \Gamma(x_0)$.

2. **upper semicontinuous in the sense of Berge** (abbr. u.s.c.-B) at x_0 if for all open sets $\Omega \supset \Gamma(x_0)$, there exists a $\delta = \delta(\Omega) > 0$ with $\Gamma(x) \subset \Omega$ for all $x \in U_\delta(x_0)$ (where $U_\delta(x_0)$ is the open neighborhood of radius δ around x_0).

3. **lower semicontinuous in the sense of Berge** (abbr. l.s.c.-B) at x_0 if for all open sets Ω with $\Omega \cap \Gamma(x_0) \neq \emptyset$, there exists a $\delta = \delta(\Omega) > 0$ with $\Omega \cap \Gamma(x) \neq \emptyset$ for all $x \in U_\delta(x_0)$.

4. **upper semicontinuous in the sense of Hausdorff** (abbr. u.s.c.-H) at x_0 if for all $\varepsilon > 0$, there exists $\delta > 0$ with $\Gamma(x) \subset U_\varepsilon(\Gamma(x_0))$ for all $x \in U_\delta(x_0)$.

3. LINEAR STOCHASTIC BILEVEL PROBLEMS

5. **lower semicontinuous in the sense of Hausdorff** (abbr. l.s.c.-H) at x_0 if for all $\varepsilon > 0$, there exists $\delta > 0$ with $\Gamma(x_0) \subset U_\varepsilon(\Gamma(x))$ for all $x \in U_\delta(x_0)$.

6. **continuous** if it is u.s.c.-H and l.s.c.-B.

7. **B-continuous** if it is u.s.c.-B and l.s.c.-B.

8. **H-continuous** if it is u.s.c.-H and l.s.c.-H.

It holds that u.s.c.-B is sufficient for u.s.c.-H, as well as l.s.c.-H is sufficient for l.s.c.-B. Thus, H-continuity and B-continuity are both stronger than the normal continuity.

Let
$$M(x) = \{y \in \mathbb{R}_+^{n_2} : Wy = h - Tx\}$$

the feasibility set of the second level. It is obvious that M is closed and convex for all x. Let $\lambda = h - Tx$ and

$$\Lambda = \{\lambda \in \mathbb{R}^{m_2} : x \in \mathbb{R}_+^{n_1}, Ax = b, \lambda = h - Tx\}.$$

Then, theorem 3.4.1. as well as corollary 3.4.1.1. from Bank et al. [5] can be applied and it holds that:

If $M(\lambda) \neq \emptyset$ for all $\lambda \in \Lambda$, then M is H-continuous on Λ.

Thus, under the assumption of complete (or sufficiently complete) recourse (A1), the feasibility set is H-continuous in $\lambda = h - Tx$ as well as in x since this is only an affine transposition.

With theorem 4.3.3. and 4.3.5.(2) from Bank et al. [5], it follows that $Y(x)$ is B-continuous:

Lemma 3.6 (can be concluded from Bank et al. [5])
Under the assumptions of (A1) and (A2), $Y(x)$ is B-continuous.

Proof:

Because of (A1), $M(x)$ is H-continuous and thus also l.s.c.-B; it is also convex and closed for all x. The set $Y(x)$ is nonempty for all x because of (A1) and bounded because of (A2). All functions are linear, thus continuous, thus lower and upper semicontinuous, as well as convex. Theorem 4.3.3. in Bank et al. [5] can then be applied to the lower level problem and it holds that $Y(x)$ is u.s.c.-B.

3.1 Linear Bilevel Problems

The only additional assumption for theorem 4.3.5.(2) in Bank et al. [5] to be applied here is that the functions have to be weakly analytic which is also the case here. So, it follows that $Y(x)$ is l.s.c.-B and corollary 4.3.5.1. in Bank et al. [5] implies that there exists a continuous function $y : \mathbb{R}^{n_1} \to \mathbb{R}^{n_2}$ with $y(x) \in Y(x)$.

Both together imply the B-continuity. □

Therefore, it can be derived from theorem 4.2.2.(2) in Bank et al. [5] that the **optimistic** value function

$$F^o(x) = c^\top x + \sum_{k=1}^N p_k \min_z \{d_k^\top z : z \in Y_k(x)\}$$

is lower semicontinuous in x under the above assumptions. So, under the assumptions of complete recourse (A1) and dual feasibility (A2) as well as suitable compactness assumptions, in the optimistic linear bilevel problem, there always exists an optimal solution.

For the **pessimistic** case, the value function can be redefined as

$$\begin{aligned} F^p(x) &= c^\top x + \sum_{k=1}^N p_k \max_z \{d_k^\top z : z \in Y_k(x)\} \\ &= c^\top x - \sum_{k=1}^N p_k \min_z \{(-d_k^\top z) : z \in Y_k(x)\}. \end{aligned}$$

This shows that the pessimistic value function is upper semicontinuous in x (due to the negativity sign).

Annotation: It does not make a difference if the leader's variable x occurs in an additive term $c_2^\top x$ in the followers objective function, i.e.,

$$Y(x) = \arg\min_{y^*} \{q^\top y^* + c_2^\top x : Wy^* + Hx = h, y^* \geq 0\}$$

because the leader's variable will always be treated as a parameter for the follower's problem. Thus, the term $c_2^\top x$ is a fixed value that the follower has to accept and cannot change. The only effect is that the value of the follower's problem rises at the amount of $c_2^\top x$. As a result, it does not have any affect on the follower's choice – which is only of interest here – and can hence be dropped. This is not any more the case if x comes in as a multiplicative term to y. Then, the follower's problem is bilinear and most of the above results do not hold any more.

3. LINEAR STOCHASTIC BILEVEL PROBLEMS

It is also notable that, in general, there actually exists a difference in the definitions of the optimistic approach. The formal definition for the optimistic bilevel problem is

$$\min_x F^o(x) \quad \text{subject to} \quad x \in X \qquad (3.2)$$

which has the same global optimum or optima as

$$\min_{x,y} F(x,y) \quad \text{subject to} \quad x \in X, \, y \in Y(x) \qquad (3.3)$$

It was shown in Dempe et al. [33] that these two definitions may differ concerning local optima. The authors there used a bilinear problem (i.e., only the value function of the second stage was bilinear in x and y, the rest was linear) to show that local optimal solutions of problem (3.2) are also locally optimal for (3.3), but not the other way around.

In the purely linear case and under the assumptions (A1) and (A2), both problems coincide due to the B-continuity of $Y(x)$.

Theorem 3.7
Under the assumptions (A1) and (A2), the linear problems (3.2) and (3.3) coincide in locally optimal solutions as well as in global optima.

Proof:
Let (x^*, y^*) be a local optimal solution to the linear problem (3.3), i.e., there exists an $\varepsilon > 0$ with $F(x^*, y^*) \leq F(x,y)$ for all $(x,y) \in U_\varepsilon(x^*, y^*)$, $x \in X$, $y \in Y(x)$. In order to show the assertion, it must hold that there exists an $\varepsilon^* > 0$ with $F^o(x^*) \leq F^o(x)$ for all $x \in U_{\varepsilon^*}(x^*)$, $x \in X$. Assume that this does not hold. Then, for all $0 < \varepsilon^*$ (without loss of generality $\varepsilon^* \leq \varepsilon$) there exists a $\bar{x} \in U_{\varepsilon^*}(x^*)$, $\bar{x} \in X$ with $F^o(x^*) > F^o(\bar{x})$. For such a $\bar{x}$ then, there exists a $\bar{y}(\bar{x}) = \arg\min_z \{d^\top z : z \in Y(\bar{x})\}$ (again without loss of generality assume that for readability the number of scenarios is $N=1$).

If it holds $\bar{y}(\bar{x}) \in U_{\varepsilon^*}(y^*) \subseteq U_\varepsilon(y^*)$, this would lead to a contradiction since $\bar{x} \in U_\varepsilon(x^*)$ – because $\varepsilon^* \leq \varepsilon$ – and thus $F(\bar{x}, \bar{y}(\bar{x})) < F(x^*, y^*)$. Therefore, $\bar{y}(\bar{x}) \notin U_{\varepsilon^*}(y^*)$ for every such $\bar{x} \in U_{\varepsilon^*}(x^*)$, $\bar{x} \in X$ with $F^o(x^*) > F^o(\bar{x})$ and every $0 < \varepsilon^* < \varepsilon$.

So, for $\varepsilon^* \to 0$ one can construct a sequence $(\bar{x}_n)$ of such variables with $\bar{x}_n \to x^*$ and $\lim_{n \to \infty} F^o(\bar{x}_n) \leq F^o(x^*)$. Because of the above result, a sequence exists for which $\arg\min_z \{d^\top z : z \in Y(\bar{x}_n)\} = \bar{y}(\bar{x}_n) \notin U_{\varepsilon^*}(y^*)$ for every n but $y^* = \arg\min_z \{d^\top z : z \in Y(\lim \bar{x}_n)\}$. Thus, in the direction of d there exists a jump in $Y(\cdot)$ which contradicts the B-continuity. $\square$

3.1 Linear Bilevel Problems

Another specialty of bilevel problems is the dependence on irrelevant constraints which was shown by Macal and Hurter [64]. In standard mathematical programs, it can normally be taken for granted that constraints which are inactive at any optimal solution could be dropped (or added) without changing the optimality of the solutions. But in bilevel programming, this is not the case!

Macal and Hurter employ a quadratic bilevel problem with no constraints and show that if an additional (nonnegativity) constraint is induced into the lower level which is not violated by the original optimum, the problem then has a much better optimal solution. Therefore, every person formulating a bilevel problem has to be very careful about the exact choice of constraints and variables.

But, again, it can be said that in the linear case, this does not hold. Theorem 1 from the mentioned paper [64] can be used to show that. In that theorem, Macal and Hurter show that a bilevel problem is independent of irrelevant constraints if and only if there exists a feasible solution to the "associated single-level problem" – which is the complete upper level joined with the lower level constraints – that is also feasible for the original bilevel problem, under the assumption that the KKT conditions for the lower level are necessary and sufficient for an optimal solution to the bilevel problem.

For linear bilevel problems, the KKT conditions possess this ability. So the theorem translated to this case would mean that a bilevel problem is independent of irrelevant constraints if there exist (x^*, y^*) solving

$$\begin{aligned} Ax &= b, \\ Wy + Tx &= h, \\ y &\geq 0 \end{aligned}$$

that also solve

$$Ax = b,$$
$$y \in Y(x) = \arg\min_y \{q^\top y : Wy = h - Tx, y \geq 0\}$$

And this is clearly the case if $Y(x)$ is bounded and nonempty, i.e., under the assumptions (A1) and (A2).

3.1.1 Existence of Solutions

Altough the definitions might in general be known, the defintions of local and global optima for bilevel problems will be provided here for completeness.

3. LINEAR STOCHASTIC BILEVEL PROBLEMS

Definition 3.8

1. A point $(x^*, y^*) \in \mathbb{R}^n \times \mathbb{R}^m$ is a **local optimistic solution** to the optimistic bilevel problem 3.2 if $y^* \in Y(x^*), x \in X$ with

$$F(x^*, y^*) \leq F(x^*, y) \quad \forall y \in Y(x^*)$$

and there exists an open neighborhood $U_\delta(x^*), \delta > 0$ with

$$F^o(x^*) \leq F^o(x) \quad \forall x \in X \cap U_\delta(x^*)$$

It is called a global optimistic solution if $\delta = \infty$ can be chosen.

2. On the other hand is $(x^*, y^*) \in \mathbb{R}^n \times \mathbb{R}^m$ a **local pessimistic solution** to the pessimistic bilevel problem

$$\min_x F^p(x) \quad \text{subject to} \quad x \in X$$

if $y^* \in Y(x^*), x \in X$ with

$$F(x^*, y^*) \geq F(x^*, y) \quad \forall y \in Y(x^*)$$

and there exists an open neighborhood $U_\delta(x^*), \delta > 0$ with

$$F^p(x^*) \leq F^p(x) \quad \forall x \in X \cap U_\delta(x^*)$$

It is called a global pessimistic solution if $\delta = \infty$ can be chosen.

3. If the lower level solution is uniquely determined for all x, then a **local optimal solution** to the general bilevel problem

$$\min_x F(x, y(x)) \quad \text{subject to} \quad x \in X$$

fulfills $y^*(x^*) \in Y(x^*), x \in X$ and there exists an open neighborhood $U_\delta(x^*), \delta > 0$ with

$$F(x^*, y^*(x^*)) \leq F(x, y(x)) \quad \forall x \in X \cap U_\delta(x^*) \text{ and } y(x) \in Y(x)$$

In many works, the authors assume for the general bilevel problem that the lower level solution is unique for every x and derive optimality conditions for such problems. For example, Dempe [28] showed that the general bilevel problem (see (1.1) on page 2) has a global optimal solution under the assumptions that

1. It has a feasible solution,

3.1 Linear Bilevel Problems

2. X is closed,

3. *Compactness* (C): The feasibility set in both variables according to the constraints of the lower level $\{(x,y) \in \mathbb{R}^n \times \mathbb{R}^m : g(x,y) \leq 0, h(x,y) = 0\}$ is nonempty and compact (here, the constraints are specialized into equality and inequality constraints), and

4. *Mangasarian-Fromovitz Constraint Qualification* (MFCQ): For all points $(x^0, y^0) \in X \times \mathbb{R}^m$ and $y^0 \in M(x^0)$ exists a direction $d \in \mathbb{R}^n$ satisfying

$$\nabla_y g_i(x^0, y^0) d < 0, \quad \text{for each } i \in \{j : g_j(x^0, y^0) = 0\}$$
$$\nabla_y h_j(x^0, y^0) d = 0, \quad \text{for each } j = 1, \ldots, q$$

where $h : \mathbb{R}^n \times \mathbb{R}^m \to \mathbb{R}^q$ and the gradients $\{\nabla_y h_j(x^0, y^0) : j = 1, \ldots, q\}$ are linearly independent.

A similar statement can be made for the optimistic bilevel problem if (MFCQ) holds for all $y \in M(x)$. Moreover, (MFCQ) could be replaced by the claim that $Y(x)$ is upper semicontinuous.

In the pessimistic case, the assumptions differ a bit. In order to have a global optimal solution for the pessimistic bilevel problem, there has to hold

1. It has a feasible solution,

2. $Y(x)$ has to be lower semicontinuous for all $x \in X$, and

3. As above, the compactness assumption (C) has to be satisfied.

This was also proven by Dempe [28].

Local optimality conditions including the *contingent* or *Bouligand cone* can be found in Dempe et al. [33]. The cone can be defined as

$$C_X(x) := \{v \in R^n : \lim_{y \to 0} \inf \frac{d_X(x+yv)}{y} = 0\}$$

where $d_X(y)$ denotes the distance from point y to the set X.

The graph of a multifunction will be $Grph(Y) = \{(x,y) \in \mathbb{R}^{n \times m} : y \in Y(x)\}$.

Lemma 3.9 (Dempe et al. [33])

If a point $(\bar{x}, \bar{y}) \in Grph(Y), \bar{x} \in X$ is a locally optimal solution of the optimistic bilevel problem, then it holds for all $(d, r) \in C_N(x, y)$, where $N = Grph(Y) \cap (X \times \mathbb{R}^m)$, that

$$\nabla F(\bar{x}, \bar{y})^\top (d, r) \geq 0.$$

3. LINEAR STOCHASTIC BILEVEL PROBLEMS

On the other hand, if $(\bar{x}, \bar{y}) \in Grph(Y), \bar{x} \in X$ and it holds

$$\nabla F(\bar{x}, \bar{y})^\top (d, r) > 0$$

for all $(d, r) \in C_N(x, y)$, then $(\bar{x}, \bar{y})$ is a locally optimal solution of the optimistic bilevel problem.

Strong regularity assumptions such as the constant rank constraint qualification (CRCQ), the strong second-order optimality condition (SSOC), and the (MFCQ) will imply the fulfillment of the assumptions in the previous lemma.

Suchlike conditions also guarantee the uniqueness of the lower level solution as well as the existence of directional derivatives, see Dempe et al. [33].

An important fact is that the constraint qualifications are dependent on the reformulation of the bilevel problem. Certain typical constraint qualifications – such as Slater or MFCQ – known from nonlinear optimization do not hold for the reformulation using the optimal value function. Further information about this property will be given in the corresponding chapters 4 and 5 on the optimal value function approach and the KKT approach, respectively.

For the linear case, it is enough to assume that the lower level is feasible, bounded, and that the lower level cost vector q is not the multiple of any row of W in order to get a unique lower level solution for all x. This is due to the fact that if the lower level is feasible and bounded, then the optimal solution will be an extreme point of the lower level feasible set or a convex combination of two (or more) suchlike points. The latter case – which results in multiple solution vectors – would only occur if the cost vector q is the multiple of a constraint row of W.

Additionally, it holds for linear bilevel problems that all optimal solutions are contained in the *constraint region*

$$M = \{(x, y) \in \mathbb{R}^n \times \mathbb{R}^m : Ax = b, \quad x \geq 0, \quad Wy = h - Tx, \quad y \geq 0\}$$

if it is nonempty and bounded.

Lemma 3.10 (see Bialas and Karwan [12] and Bard [6])
The solution of the linear bilevel problem occurs at a vertex of M, if M is nonempty and compact.

It follows that the linear bilevel problem always has an optimal solution regardless whether it is optimistic or pessimistic, if M is nonempty and bounded.

3.1 Linear Bilevel Problems

3.1.2 Reformulations

As traditional solvers widely fail with bilevel problems, a hand full of techniques were developed in order to be able to solve these kind of problems efficiently. It is worth mentioning that optimality conditions derived for each approach are not related, in general. Different optimality conditions hold in different approaches.

Using the *Karush Kuhn Tucker conditions* (abbr. KKT conditions), in order to replace the lower optimization problem $Y(x)$ by constraints that are necessary and sufficient for an optimal solution of such, was maybe the first but is certainly the most frequent approach. Therefore, the lower level has to be convex in the lower level variable and some constraint qualification has to be satisfied, e.g.:

- *Linearity constraint qualification*: the components of the constraint functions are affine (which is abbreviated LCQ),

- *Linear independence constraint qualification*: the gradients of the active constraints are linearly independent (abbr. LICQ),

- *Mangasarian-Fromovitz constraint qualification*: the gradients of the active inequality constraints and the gradients of the equality constraints are positive-linearly independent (MFCQ),

- *Constant rank constraint qualification*: for each subset of the gradients of the active inequality constraints and the gradients of the equality constraints the rank at a vicinity of a solution is constant (CRCQ),

- *Slater constraint qualification* (SCQ): there exists a point such that all equality constraints are satisfied and all inequality constraints are strictly fulfilled (i.e., unequal zero),

among others. In the nonaffine case, it is assumed that the objective and the constraint functions are continuously differentiable.

But, due to the fact that new variables are introduced through the KKT conditions, the original and the reformulated problem only coincide in global optima. Additionally, the remaining problem is nonconvex due to the complementarity constraints. A thorough investigation on that matter will be presented in chapter 5.

3. LINEAR STOCHASTIC BILEVEL PROBLEMS

Another approach is to use the *optimal value function of the lower level*. To do so, the (optimistic) upper level problem is extended by the constraints of the lower level problem plus an additional constraint that forces the lower level objective costs to be smaller or equal than that optimum (for the minimization problem). In the linear case, that would mean

$$q^\top y \leq \phi(x) := \phi(q, h - Tx)$$

is added to the problem. The reformulated problem is linear except for that constraint which contains (at best) a piecewise linear function. More information on that matter will be given in chapter 4.

A third approach, that has also gained some interest in the last decades, is using the *normal cone* of the lower level problem. Since no chapter is devoted to this topic, it will be presented here in more detail. Dempe and Zemkoho [37] have investigated this topic and the definitions as well as properties are taken from that paper.

If for the general bilevel problem

$$\begin{aligned} & \min \quad F(x,y) \\ & \text{s.t.} \quad G(x) \leq 0 \\ & \quad y \in Y(x) = \arg\min_{y^*}\{f(x, y^*) : g(x, y^*) \leq 0\} \end{aligned}$$

it is assumed that the lower level problem is convex in y for all x with $G(x) \leq 0$, then the lower level solution set takes the form

$$Y(x) = \{y \in \mathbb{R}^m : 0 \in \nabla_y f(x,y) + N_{K(x)}(y)\} \tag{3.4}$$

where $K(x) = \{y^* \in \mathbb{R}^m : g(x, y^*) \leq 0\}$ is the lower level feasibility set, ∇_y is the vector differential operator and

$$N_S(y) = \{v \in \mathbb{R}^m : \langle v, u - y \rangle \leq 0, \forall u \in S\}$$

for $y \in S$ defines the normal cone to a convex set $S \subseteq \mathbb{R}^m$. The reformulated problem hence belongs to the class of *optimization problems with variational inequality constraint(s)*.

This reformulation may be seen as a compact form of KKT conditions, but has the advantage that not only global but also local solutions coincide with the original formulation, see theorem 3.2 in Dempe and Zemkoho [37]. But, this nice property is lost as soon as the normal cone has to be computed since the local optimality has to

3.1 Linear Bilevel Problems

be checked for all $u \in K(x)$; thus, this approach is mostly interesting for theoretical issues.

Other computational approaches to solve bilevel problems are:

1. *Extreme Point Approaches:* These are mostly applied to linear versions. In the linear case, these algorithms use the fact that the *relaxed feasible region*

$$\mathcal{R} := \{x \in \mathbb{R}^n, y \in \mathbb{R}^m : G(x) \leq 0, g(x,y) \leq 0\}$$

contains the optimal solution if it is nonempty. Thus, bases were scaned on different criteria by Candler and Townsley [19] as well as Bialas and Karwan [12], among others.

2. *Descent Methods:* These methods can only be used in the case when the lower level solution is unique, thus, $y(x)$ is uniquely determined. Starting from a feasible point x, a feasible direction d is found such that $(x+d, y(x+d))$ is bilevel feasible and decreases the upper level costs. But, in the bilevel case it is not so straightforward to find such a descent direction. Attempts have been made by Kolstad and Lasdon [58], Savard and Gauvin [79], and Vicente et al. [88].

3. *Penalty Function Methods:* Here, the lower level is replaced by a penalty-term either in the upper level objective function or said constraints. A weighted sum of the lower level objective function and constraints is used to define the lower level variable y and an additional scalar (or vector). Normally, it is used for nonlinear bilevel problems as has been done by Aiyoshi and Shimizu [1], Ishizuka and Aiyoshi [54], and Case [21].

4. *Trust Region Methods:* Developed for nonlinear bilevel problems, these methods again start with a feasible solution (x,y) and solve a subproblem that represents a linear approximation of the inherent objective and constraint functions at that point (x,y). The new optimal solution is tested for its ability to be the next iterate and the radius for which the functions are linearized. This is repeated until convergence occurs. Examples for that method can be found in Colson et al. [25] or Dempe and Bard [30].

More detailed information about different methods can be found in Bard [7], Dempe [28], or Colson et al. [26].

3. LINEAR STOCHASTIC BILEVEL PROBLEMS

3.2 Under Uncertainty

Two approaches will be exploited to analyze bilevel problems under uncertainty, the computational and the theoretical. For the computational perspective, in most cases, the uncertainty in the model should permit to be discretized or properly evaluated at certain points in order to compute a solution in finite time. Theoretically, all kinds of risk functionals and distributions can be expected.

The optimistic stochastic programming extension of a linear bilevel problem reads as

$$\begin{align}
\min \quad & \mathcal{Q}_\omega[F(x, y(\omega))] = \mathcal{Q}_\omega[c^\top x + d^\top(\omega)y(\omega)] \\
\text{s.t.} \quad & Ax = b, \, x \geq 0 \\
& y(\omega) \in Y(x, \omega), \quad \text{for almost all } \omega \in \Omega \\
\text{where} \quad & Y(x, \omega) = \arg\min_{y^*}\{q^\top(\omega)y^* : W(\omega)y^* + H(\omega)x = h(\omega), \, y^* \geq 0\}.
\end{align} \quad (3.5)$$

$(\Omega, \mathcal{A}, \mathcal{P})$ is a probability space where Ω is a discrete or continuous set of outcomes, $\mathcal{A}$ a σ-algebra, $\mathcal{P}$ a probability measure, and $\mathcal{Q}_\omega$ some risk functional on that space.

Although this thesis focuses on the computational viewpoint, this section starts with a brief review of some interesting theoretical facts, i.e., when the set Ω of outcomes is continuous.

$\mathcal{Q}_\omega$ can be any risk functional. Note that every lower level parameter is provided with randomness. Thus, this presents a general form of a linear stochastic bilevel problem. Other kind of stochastic problems exist where e.g. the lower level is a two-stage stochastic problem in itself, but, this is either not linear anymore or may rather be a multilevel problem instead of bilevel.

If the leader is interested only in the expected best value on average of the above problem, then, $\mathcal{Q}_\omega = \mathcal{Q}_E$ and

$$\mathcal{Q}_E[F(x, y(\omega))] = c^\top x + \int_\Omega d^\top(\omega) y(\omega) \, dF(\omega)$$

where F denotes the cummulative distribution function, assumed to be continuously differentiable with $\int_\Omega dF(\omega) = 1$.

3.2 Under Uncertainty

Corollary 3.11 (Patriksson and Wynter [71])
The graph $Grph(Y)^1$ is closed if

1. For each $\omega \in \Omega$ and each $x \in X = \{x \in \mathbb{R}^n : Ax = b, x \geq 0\}$, it holds that $Y(x, \omega) \neq \emptyset$,

2. the follower's data influenced by randomness, q, W, H and h, are continuous on Ω.

The expected-value based stochastic linear bilevel problem (3.5) with an infinite set of outcomes has at least one optimal solution if, additionally to 1. and 2., it holds that

3. there exists an $(x, y(\omega)) \in Z(\omega) = \{(s, t(\omega)) \in Grph(Y(\omega)) : s \in X\}^2$ in every $Z(\omega), \omega \in \Omega$ for some x, and

4. $f(x, y(\omega), \omega) = c^\top x + d^\top(\omega) y(\omega)$ has bounded lower level sets on $\cup_{\omega \in \Omega} Z(\omega)$, i.e.,

$$L_c^-(f) = \bigcup_{\omega \in \Omega} \{(x, y(\omega)) \in Z(\omega) : f(x, y(\omega), \omega) \leq c\}$$

are bounded for all scalars $c \in \mathbb{R}$.

Actually, this corollary was shown for the general (nonlinear) case, but is simplified for the present linear case.

Equivalently as done for two-stage problems, other risk functionals can be applied to the above problem (3.5), e.g. the excess probability

$$\mathcal{Q}_\omega = Q_\mathbb{E}[F(x, y(\omega))] + \mathcal{P}(\{\omega \in \Omega : f(x, y(\omega), \omega) > \eta\}),$$

the Value-at-Risk

$$\mathcal{Q}_\omega = Q_\mathbb{E}[F(x, y(\omega))] + Q_{\alpha VaR}(x) := \inf\{\eta : \mathcal{P}(\{\omega \in \Omega : f(x, y(\omega), \omega) \leq \eta\}) \geq \alpha\},$$

or the Conditional Value-at-Risk

$$\mathcal{Q}_\omega = Q_\mathbb{E}[F(x, y(\omega))] + Q_\alpha(x, \eta) := \frac{1}{\alpha} \int_0^\alpha G^{-1}(u) du.$$

Ivanov [55] analyzed continuity properties for the Value-at-Risk.

---
[1] $Grph(Y) = \{(x, y(\omega), \omega) \in \mathbb{R}^{n+m} \times \Omega : y(\omega) \in Y(x, \omega)\}$
[2] $Grph(Y(\omega)) = \{(x, y(\omega)) \in \mathbb{R}^{n+m} : y(\omega) \in Y(x, \omega)\}$

3. LINEAR STOCHASTIC BILEVEL PROBLEMS

To the author's knowledge, other existence results for these risk functionals have not been addressed, yet.

In practice, it is often difficult to evaluate the risk functional in (3.5) for problems of realistic size due to the difficulty in calculating the multiple integrals. Therefore, it is convenient to assume that Ω is discrete and finite, i.e., $\Omega = \{1, \ldots, N\}$.

In the case of the expected value, the evaluation function Q_E then becomes

$$Q_E[F(x, y(\omega))] = c^\top x + \sum_{\omega=1}^{N} \pi(\omega) d(\omega)^\top y_\omega$$

where $\sum \pi(\omega) = 1$ and $0 \leq \pi(\omega) \leq 1$ for $\omega = 1, \ldots, N$. The problem then becomes

$$\begin{aligned} \min \quad & c^\top x + \sum_{\omega=1}^{N} \pi(\omega) d(\omega)^\top y_\omega \\ \text{s.t.} \quad & Ax = b, \, x \geq 0 \\ & y_\omega \in Y(x, \omega), \quad \omega = 1, \ldots, N \end{aligned} \quad (3.6)$$

where $Y(x, \omega) = \arg\min_{y^*}\{q^\top(\omega) y^* : W(\omega) y^* + H(\omega) x = h(\omega), \, y^* \geq 0\}$.

This problem can also be called the deterministic equivalent formulation.

Corollary 3.11 can be similarly applied to problem (3.6). The linear stochastic bilevel problem referred to in the following chapters correspond to this problem (3.6) with the adjustment that the lower level matrices are not dependent on the randomness, i.e., $W(\omega) \equiv W$ and $H(\omega) \equiv H$.

4

Optimal Value Function Approach

In the present chapter, the potential of the optimal-value-function approach for its expansion to the stochastic case is explored. In particular, the deterministic method for quasiconcave minimization due to Tuy et al. [85] is examined in more detail, leading to the outcome that there is no direct way from the deterministic to the stochastic setting when it comes to investigating possible decomposition techniques.

Under the assumption that the general optimistic bilevel problem 1.1 has a feasible solution, its reformulation using the optimal value function of the lower level is

$$
\begin{aligned}
\min \quad & F(x,y) \\
\text{s.t.} \quad & G(x) \leq 0 \\
& g(x,y) \leq 0 \\
& f(x,y) - \phi(x) \leq 0
\end{aligned}
\tag{4.1}
$$

where $\phi(x) = \min_y \{f(x,y) : g(x,y) \leq 0\}$. It is $\phi(x) = +\infty$ if $Y(x) = \emptyset$ for convenience, as well as $X = \{x : G(x) \leq 0\}$, $K(x) = \{y \in \mathbb{R}^m : g(x,y) \leq 0\}$ the upper, resp. lower level constraint set.

To ensure that $\phi(x)$ is finite for all x, it is assumend here that $Y(x)$ is nonempty and compact.

It holds that the original optimistic bilevel problem and this reformulation 4.1 have the same local and global optima. This is due to the fact that any parametric opti-

4. OPTIMAL VALUE FUNCTION APPROACH

mization program, e.g.

$$\min_y H(x,y), \quad \text{s.t. } C(x,y) \leq 0$$

can be rewritten as

$$H(x,y) \leq H(x,t), \, \forall t \text{ with } C(x,t) \leq 0, \quad \text{and } C(x,y) \leq 0$$

and with $H(x) = \min_t \{H(x,t) : C(x,t) \leq 0\}$ has the same feasible and optimal solutions as

$$H(x,y) \leq H(x), \quad \text{s.t. } C(x,y) \leq 0$$

But despite that nice property, the drawback here is due to the nature of the optimal value function itself, which can be nonsmooth. Besides, some constraint qualifications as the Mangasarian-Fromovitz constraint qualification, the linear independence constraint qualification, and the Slater condition do not hold in this case which was shown by Ye and Zhu [95], [96] in terms of Clarke's generalized subdifferential. Dempe and Zemkoho [35] showed that the Mangasarian-Fromovitz constraint qualification does also not hold if Mordukhovich's subdifferential is used.

Instead, other constraint qualifications have been shown to work in this case and have been used to find algorithmic solutions. Especially the concept of *partial calmness* has been proven to work very well for bilevel problems. Partial calmness is attained at a point $(\bar{x}, \bar{y})$ of bilevel problem (4.1) if and only if there exist $\alpha > 0$ and a neighborhood U of $(\bar{x}, \bar{y}, 0) \in \mathbb{R}^n \times \mathbb{R}^m \times \mathbb{R}$, such that:

$$F(x,y) - F(\bar{x}, \bar{y}) + \alpha |u| \geq 0, \tag{4.2}$$

for all $(x, y, u) \in U$ fulfilling $f(x,y) - \phi(x) + u = 0$ and $(x,y) \in M$, where M was the joint set of lower and upper level constraints. Not surprisingly, partial calmness is a necessary and sufficient criterion for the exactness of the penalty function approach

$$\min F(x,y) + \lambda(f(x,y) - \phi(x)), \quad \text{s.t. } x \in X, \, y \in K(x).$$

The concept was first used on bilevel problems by Ye and Zhu [95] where they derived necessary optimality conditions for different kinds of bilevel problems. In particular, it can be shown that purely linear bilevel problems are always partially calm if they are solvable (using their proposition 5.1 and the work of Burke and Ferris [17]). Dempe and

Zemkoho strengthened that result to models where the lower level is bilinear, the upper level cost function F is Lipschitz continuous, the domain of the lower level $dom Y = \mathbb{R}^n$ is defined everywhere and no upper level constraints are imposed. In fact, the partial calmness depends strongly on the cost function F, on the upper level constraint set X and, of course, on the structure of the lower level problem. But, there also exist models where the special structure of the lower level problem produces a partially calm problem independent of the upper level functions, as was shown by Dempe et al. [32].

For the more general case of partial calmness, Ye and Zhu [95] proposed the concept of *uniformly weak sharp minima*. The family of parametric optimization problems $\{Y(x) : x \in X\}$ has a uniformly weak sharp minimum if there exists $\mu > 0$ such that

$$f(x,y) - \phi(x) \geq \mu \, d(y, Y(x)), \quad \forall y \in K(x) \text{ and } x \in X$$

where $d(y, Y)$ is the euclidean distance of point y to set Y.

The authors then showed that a bilevel problem is partially calm at a local solution if the upper level cost function F is locally Lipschitz continuous and the family of parametric optimization problems $\{Y(x) : x \in X\}$ has a uniformly weak sharp minimum.

Additionally, Henrion and Surowiec [49] found out that the calmness (which can be related to some Lipschitz-like behaviour) of the set-valued mapping

$$\mathcal{M}(v) = \{(x,y) \in M : f(x,y) - \phi(x) \leq v\}$$

– called value function constraint qualification, short VFCQ – could also be used as a sufficient condition for partial calmness if F is locally Lipschitz continuous. VFCQ is necessary for the existence of a uniformly weak sharp minimum. These implications are not trivial since the authors provide an example that shows that the VFCQ is strictly weaker than the existence of a uniformly weak sharp minimum. Additionally, they showed that there are problems where partial calmness does not hold at any local solution, but constraint qualifications using the KKT approach worked in that case.

Despite that, also other constraint qualification have been found to work for certain classes of bilevel problems. Ye [94] showed that the nondifferentiable Arrow-Hurwicz-Uzawa CQ, the generalized Zangwill CQ, the nondifferentiable Kuhn-Tucker CQ, and

4. OPTIMAL VALUE FUNCTION APPROACH

the nondifferentiable Abadie CQ are all applicable constraint qualifications for nonlinear bilevel problems. She used these qualifications along with the Michel-Penot subdifferential in order to derive KKT type optimality conditions.

Dempe and Zemkoho [35] used the generalized differentiation theory of Mordukhovich, rephrased a dual form of the MFCQ – which was dualized by Ye and Zhu [95] – and weakened the constraint qualification in order to get a new optimality condition that holds for a broad class of bilevel problems.

These papers also use the inner semicontinuity and the inner semicompactness of the optimal solution function Y in order to characterize the behaviour of said multifunction. *Inner semicompactness* holds for Y at $\bar{x}$ if and only if for every sequence $x_k \to \bar{x}$ with $Y(x_k) \neq \emptyset$, there exists a sequence $y_k \in Y(x_k)$ that contains a convergent subsequence.

Y is *inner semicontinuous* at $(\bar{x}, \bar{y})$ if for every sequence $x_k \to \bar{x}$ there is a sequence $y_k \in Y(x_k)$ that converges to $\bar{y}$. Clearly, inner semicontinuity is sufficient for inner semicompactness. But it is also necessary for the following *Aubin (or Lipschitz-like) property* which holds at a point $(\bar{x}, \bar{y})$ if there are neighborhoods U of $\bar{x}$ and V of $\bar{y}$ as well as a constant $L > 0$ such that

$$d(y, Y(x_2)) \leq L\|x_1 - x_2\|$$

for all $x_1, x_2 \in U$ and all $y \in Y(x_1) \cap V$.

The inner semicompactness of Y at $\bar{x}$ together with K satisfying the Aubin property at $(\bar{x}, y)$ for all $y \in Y(\bar{x})$ imply the Lipschitz continuity of φ around $\bar{x}$ (see Mordukhovich [65]). φ is also Lipschitz continuous if Y is inner semicontinuous (see Mordukhovich and Nam [66]). For example, Dempe and Zemkoho [35] use this together with the above mentioned weak MFCQ condition in order to derive KKT-type and other necessary optimality conditions.

Also, Dempe et al. [32] used the inner semicompactness and inner semicontinuity of Y together with partial calmness and some regularity assumptions for the lower and upper level in order to derive KKT-type and Fritz-John-type necessary optimality conditions. Besides, they provided some cases in which Y is inner semicontinuous, e.g. if $Y(x)$ is a singleton, if the lower level constraint functions are weakly analytic as in Definition 3.4, or if the lower level is linear in both variables.

4.1 A Method for Linear Bilevel Programs

This section will illustrate the difficulties that occur when following this approach.

So, the linear deterministic equivalent of the linear stochastic bilevel problem is, as presented in section 3.2,

$$\begin{aligned}
\min_{x,y} \quad & c^\top x + \sum_{k=1}^{N} d_k^\top y_k \\
\text{s.t.} \quad & Ax = b \\
& Wy_k + Hx = h_k, \quad k = 1, \ldots, N \\
& q_k^\top y_k \leq \phi_k(x), \quad k = 1, \ldots, N \\
& x, y_k \geq 0, \quad k = 1, \ldots, N
\end{aligned} \quad (4.3)$$

where $\phi_k(x)$ is the optimal value function of the lower level of scenario k and $y = (y^1, \ldots, y^N)$. It has inner semicontinuous multifunctions Y_k, along with multiple constraints that have one variable in common: x.

As stated in section 2.1, these functions ϕ_k are piecewise linear and convex. In particular, there exist $\delta_{j,k}$, $j = 1, \ldots, J$ – which can be associated with vertices of the dual problem's feasible set – such that

$$\phi_k(x) = \max_{j=1,\ldots,J} \left[\delta_{j,k}^\top (h_k - Hx) \right]. \quad (4.4)$$

The task now is to find a method that allows for decomposition into single scenarios. The biggest problem lies in the fact that each scenario needs the same leader variable x in order to give the correct solution to the upper level.

In the following, the method of Tuy et al. [85] will be exploited and enhanced to the case of multiple scenarios. In their paper, a solution method for linear bilevel problems is presented that is based on a branch and bound technique. In short summary, the authors use the optimal value reformulation, which they call a linear program with an additional reverse convex constraint, and formulate an equivalent quasiconcave minimization problem using the polar of a set C. Since the polar C^* is hard to compute, an outer approximation scheme is used in order to subdivide a bigger set K^* into cones for which lower and upper bounds are evaluated.

Therefore, denote M the joint set of lower and upper level constraints as well as

$$C_k := \left\{ (x, y) \ : \ \phi_k(x) \leq q_k^\top y_k \right\}$$

4. OPTIMAL VALUE FUNCTION APPROACH

Note that the inequality is stated the other way round compared to model (4.3). Because of the above observation (4.4), it holds that

$$C_k = \{(x,y) : \delta_{j,k}^\top Hx + q_k^\top y_k \geq \delta_{j,k}^\top h_k\,,\, j = 1,\ldots,J\}, \qquad (4.5)$$

since $\phi_k(x) \leq q_k^\top y_k$ implies $\max_{j=1,\ldots,J}\left[\delta_{j,k}^\top(h_k - Hx)\right] \leq q_k^\top y_k$ which is equivalent to the above representation (4.5). This shows that C_k is a polyhedron whose interior is given by

$$\begin{aligned}int\, C_k &= \{(x,y) : \delta_{j,k}^\top Hx + q_k^\top y_k > \delta_{j,k}^\top h_k\,,\, j = 1,\ldots,J\} \\ &= \{(x,y) : \phi_k(x) < q_k^\top y_k\}\end{aligned}$$

If the linear problem

$$\min\left\{c^\top x + \sum_{k=1}^N d_k^\top y_k : (x,y) \in M\right\} \qquad (4.6)$$

has an optimal solution $(\bar{x},\bar{y})$ that is already feasible for problem (4.3), then, the optimal solution of the original bilevel problem is found. But, if it is not optimal for at least one second level, i.e., it holds $(\bar{x},\bar{y}) \in int\, C_k$ for one k (or more), then, some cuts would have to be added to problem (4.6) in order to find the optimal bilevel solution.

Now, if

$$C := \bigcup_{k=1}^N C_k,$$

then $\bigcup_{k=1}^N int\, C_k \subseteq int\, C$ because the interior of a set is the union of all open sets contained in it, $int\, C_k \subseteq C$, and $int\, C_k$ is open for all k.

So, for the above suboptimal point, it holds $(\bar{x},\bar{y}) \in int\, C_k \subseteq int\, C$. The use of the Minkowski difference entails that two new sets can be defined:

$$\hat{C} := C - (\bar{x},\bar{y}) = \bigcup_{k=1}^N \underbrace{\{C_k - (\bar{x},\bar{y})\}}_{=:\hat{C}_k},$$

$$\hat{D} := D - (\bar{x},\bar{y}).$$

Lemma 4.1

For $k = 1,\ldots,N$ it holds that

$$K_k := \{(x,y) : Hx \leq 0\,,\, q_k^\top y_k \geq 0\} \subseteq \hat{C}_k.$$

4.1 A Method for Linear Bilevel Programs

Proof:
It is to show that $K_k + (\bar{x}, \bar{y}) \subseteq C_k$. So, if $(x,y) \in K_k$, then the assertion amounts to verifying that $(\bar{x} + x, \bar{y} + y) \in C_k$.

First, due to Farkas' Lemma, notice that for arbitrary y' fulfilling

$$H\bar{x} + Wy'_k \leq h_k, \ y'_k \geq 0$$

it holds

$$H(\bar{x} + x) + Wy'_k \leq h_k, \ y'_k \geq 0$$

which yields

$$\phi_k(\bar{x} + x) \leq \phi_k(x).$$

Thus,

$$\phi_k(\bar{x} + x) \leq \phi_k(x) \leq q_k^\top \bar{y}_k \leq q_k^\top (\bar{y}_k + y)$$

due to the definition of the points $(\bar{x}, \bar{y})$ and (x, y). This provides the assertion. $\square$

It follows that

$$K := \bigcup_{k=1}^{N} K_k \subseteq \bigcup_{k=1}^{N} \hat{C}_k = \hat{C}$$

where K is a finite union of polyhedral cones.

Using (4.5), the following representation holds true

$$\begin{aligned}\hat{C}_k &= \{(x,y) : \delta_{j,k}^\top H(x + \bar{x}) + q_k^\top (y_k + \bar{y}_k) \geq \delta_{j,k}^\top h_k, \ j = 1, \ldots, J\} \\ &= \{(x,y) : \delta_{j,k}^\top Hx + q_k^\top y_k \geq \underbrace{\delta_{j,k}^\top (h_k - H\bar{x}) - q_k^\top \bar{y}_k}_{\leq 0}, \ j = 1, \ldots, J\}\end{aligned}$$

Recall the definition of the *polar* V^* to a set V is defined to be

$$V^* := \{u : u^\top v \leq 1, \forall v \in V\}$$

and it holds that

$$\hat{C}^* = \bigcap_{k=1}^{N} \hat{C}_k^* \subseteq \bigcap_{k=1}^{N} K_k^* = K^* \tag{4.7}$$

According to Schrijver [80], Theorem 9.1(iv), it holds that

$$K_k^* = cone\{H_1^\top, \ldots, H_{m_2}^\top\} \times \{0_1\} \times \ldots \times \{0_{k-1}\} \times cone\{-q_k\} \times \{0_{k+1}\} \times \ldots \times \{0_N\}$$

where *cone* defines the conical hull of the containing vectors and H_i is the i-th row of H.

4. OPTIMAL VALUE FUNCTION APPROACH

Define
$$u = \begin{pmatrix} x \\ y \end{pmatrix}, \quad \gamma^\top u = c^\top x + \sum_{k=1}^{N} d_k^\top y_k$$

and the function $f : K^* \to (-\infty, \infty]$ as
$$f(v) := \inf\{\gamma^\top u : u \in \hat{D}, v^\top u \geq 1\}$$

with the usual convention $\inf \emptyset = \infty$.

Next, Tuy et al. [85] prove the following proposition which is crucial for their algorithm, but does **not** hold here.

Proposition 4.2 (Tuy et al. [85])
Problem
$$\min_u \{\gamma^\top u : u \in \hat{D} \setminus int\ \hat{C}\} \tag{4.8}$$

is equivalent to the original problem (4.3) and to the following
$$\min\{f(v) : v \in \hat{C}^*\} \tag{4.9}$$

in the sense that the optimal values coincide, and, if $\bar{v}$ solves (4.9), then
$$\bar{u} \in argmin\,\{\gamma^\top u : u \in \hat{D}, v^\top u \geq 1\}$$

solves (4.8).

The first equivalence is actually derived in the text on page 246 in Tuy et al. [85]. In their case, it holds true that
$$int\ C = \{u : \phi(x) < q^\top y\}.$$

But, for linear stochastic bilevel problems with C defined as above, this equivalence does not hold. To prove that, two cases have to be distinguished. First, assume that the problem has complete recourse $(h_k - T_k x \in posW := \{t : Wy = t, y \geq 0\}$ for all k and $x \in \mathbb{R}^n$) and, for simplicity, assume that there exist only two scenarios, thus

$$C = \underbrace{\{(x, y_1, y_2) \in \mathbb{R}^{n+2*m} : q_1^\top y_1 \geq \phi_1(x)\}}_{=C_1} \cup \underbrace{\{(x, y_1, y_2) \in \mathbb{R}^{n+2*m} : q_2^\top y_1 \geq \phi_2(x)\}}_{=C_2}$$

Because of the complete recourse, there is a vector $y_i, i = 1, 2$ for every x such that $\phi_i(x) = q_i^\top y_i$ or $\phi_i(x) = -\infty$ is unbounded. In both cases, the set C_1 (resp. C_2) consists of something like $\mathbb{R}^n \times I_1 \times \mathbb{R}^m$ (resp. $\mathbb{R}^n \times \mathbb{R}^m \times I_2$) where $I_i \subseteq \mathbb{R}^m, i = 1, 2$

4.1 A Method for Linear Bilevel Programs

since each set only restricts one of the scenario variables. Thus, $C = \mathbb{R}^n \times \mathbb{R}^{2*m}$ as it is the union of both sets. So, it holds that $C = \mathbb{R}^n \times \mathbb{R}^{2*m} = \hat{C} = int\, C$ (as well as $C^* = 0$) and the set $\hat{D} \setminus int\, \hat{C} = \emptyset$. This would mean that problem (4.8) is in this case equal to $\min\{\emptyset\} = +\infty$. Thus, the first equivalence does not hold for linear stochastic bilevel problems with complete recourse.

For the case of incomplete recourse, the above equality $int\, C = \{u : \phi(x) < q^\top y\}$ does not hold which will be shown by the next example.

Example 4.3
Again, two simple scenarios are assumed. Set

$$\phi_1(x) = \min\{-5y_1 : y_1 \leq x, y_1 \geq 2\}$$

and

$$\phi_2(x) = \min\{-5y_2 : y_2 \leq x, y_2 \geq 0\}.$$

Then,

$$\phi_1(x) = \begin{cases} -5x & , x \geq 2 \\ +\infty & , x < 2 \end{cases} \quad \text{and} \quad \phi_2(x) = \begin{cases} -5x & , x \geq 0 \\ +\infty & , x < 0 \end{cases}$$

According to the definition, the set C consists of

$$\begin{aligned} C &= \{(x, y_1, y_2) : -5y_1 \geq -5x, x \geq 2\} \cup \{(x, y_1, y_2) : -5y_2 \geq -5x, x \geq 0\} \\ &= \{(x, y_1, y_2) : y_1 \leq x, x \geq 2\} \cup \{(x, y_1, y_2) : y_2 \leq x, x \geq 0\} \\ &= [2, \infty) \times \mathbb{R} \times \mathbb{R} \cup [0, \infty) \times \mathbb{R} \times \mathbb{R} \\ &= [0, \infty) \times \mathbb{R}^2 \end{aligned}$$

and $int\, C = (0, \infty) \times \mathbb{R}^2$, but for $x = 1$ it is $\phi_1(1) = +\infty \not< -5y_1, y_1 \in \mathbb{R}$.

Additionally, if no complete recourse can be assumed, the set C may not be convex in x, if x is a vector of at least two entries (because the union of two convex sets does not have to be convex).

An important step for the second equivalence in the above theorem is this equality

$$int\, \hat{C} = \{u : u^\top v < 1, \forall v \in \hat{C}^*\}.$$

But, as C does not have to be convex, it follows that this equivalence does not hold in the case of incomplete recourse (in the case of complete recourse, this equation does hold trivially since $\hat{C}^* = \{0\}$).

4. OPTIMAL VALUE FUNCTION APPROACH

4.2 Alternatives

The example and discussion showed that the union of the sets C_k does not enable decomposition as for instance possible in stochastic programming. This is due the fact that each set C_k uses the vectors x **and** y_k, but does not limit other scenarios. An opportunity would be to define C differently, for example

$$C := \underbrace{\{x \in \mathbb{R}^n : Ax = b,\, x \geq 0\}}_{=:X} \times \underset{k=1}{\overset{N}{\times}} \underbrace{\left\{y_k \in \mathbb{R}^m : \exists x \in X \text{ with } \phi_k(x) \leq q_k^\top y_k\right\}}_{=:C_k}$$

But then, the relation with the sets K and K_k would get lost since they depend on both variables x and y, but the sets C_k only depend on their scenario variable. And the polar of the sets K and K_k become a crucial instrument in the later branching (the set K^* is subdivided in order to find the boundaries of C^*). Additionally, in order to be in the interior of C, a point $(\bar{x}, \bar{y})$ would have to be suboptimal for every single scenario, thus, cutting off points which are already "partially optimal".

To the best knowledge of the author, most of the other algorithmic approaches for linear bilevel problems either use the Karush-Kuhn-Tucker conditions or are somehow a vertex enumeration approach – due to the fact that an optimal solution of the bilevel problem occurs at a vertex of the relaxed feasible set M.

To use decomposition methods as presented in chapter 2 directly on the problem (4.3) would also not work since relevant value functions are concave rather than convex.

Another possibility would be to use the algorithms proposed for nonlinear bilevel problems and check those for their ability to use them on linear stochastic bilevel problems (and allowance of decomposition methods). That might be a part of future research.

5

Karush-Kuhn-Tucker Approach

The Karush-Kuhn-Tucker (abbr. KKT) conditions are a very intuitive way to replace convex lower level problems and have been widely used in the industry to solve bilevel programs. At first in this chapter, a general introduction will be given on the subject of KKT conditions and their utilization in bilevel programming. It will be followed by a section that covers the properties of the reformulated bilevel problems. The next chapter 6, then, will present an algorithm that is based on this approach.

5.1 Properties in the General Case

For the general optimistic bilevel problem, the KKT approach would be to reformulate the problem as an one-level problem of the following type

$$\begin{aligned}
\min_{x,y} \quad & F(x,y) \\
\text{s.t.} \quad & G(x,y) \leq 0 \\
& g(x,y) \leq 0 \\
& u \geq 0 \\
& \nabla_y f(x,y) + u^\top \nabla_y g(x,y) = 0 \\
& u^\top g(x,y) = 0
\end{aligned} \quad (5.1)$$

where ∇ is the gradient operator. Beside the convexity, the lower level problem has to fulfill some regularity condition, such as one of those stated in section 3.1.2, in order to be reformulated as above. Due to the last constraint in (5.1), these problems are also called mathematical problems with complementarity constraint (MPCC).

5. KARUSH-KUHN-TUCKER APPROACH

If the lower level is nonconvex, the above problem (5.1) would describe a feasible set that is larger than the solution set of the original bilevel problem and, thus, could just be used as a (vague) bound. On the other hand, if regularity conditions were missing, the MPCC might not even have a global optimal solution although the bilevel problem has and even if the feasible set M is not empty and bounded – this was shown through a quadratic example by Dempe and Dutta [31].

Also for this approach, several constraint qualifications do not hold. Flegel [41] showed in his dissertation that for MPCCs the constraint qualification (MFCQ) does not hold (at any feasible point). From this, it follows that also the (LICQ) does not hold since it is sufficient for (MFCQ). Additionally, it can be easily verified that the Slater constraint qualification does not hold because there is no point such that

$$u > 0, g(x,y) < 0 \quad \text{and} \quad u^\top g(x,y) = 0.$$

Instead, he showed that the Guignard constraint qualification can be applied directly to the above problem deriving M(ordukhovich)-stationary points.

This constraint qualification was pursued by Dempe and Zemkoho [36]. They showed that a local optimal solution of a KKT reformulated bilevel problem satisfying the Guignard CQ also fulfills S(trong)-stationary conditions. But more important, they considered the so called basic constraint qualification, which is a generalization of the dual form of the MFCQ using Mordukhovich's normal cone and subdifferential. The authors use this basic CQ on an optimization problem with operator constraint, i.e., the constraint is of the form $z \in \Omega \cap \phi^{-1}(\Lambda)$ for some locally Lipschitz-continuous multifunction ϕ and closed sets $\Omega \subseteq \mathbb{R}^l$ and $\Lambda \subseteq \mathbb{R}^m$. They show that local optimal solutions correspond for certain representations in order to derive M- and S-type stationarity conditions for local optimal solutions of the KKT reformulation. Maybe interesting is the fact that the parameters of the stationarity conditions are all bounded.

In addition to hold for the original bilevel problem, some regularity assumption has to be fulfilled by the lower level and the basic CQ has to hold for all Lagrangian multipliers at a point $(\bar{x}, \bar{y})$. The authors also use the concept of partial calmness, normally used for the optimal value function approach, see (4.2) on page 44, in order to derive the S-type optimality conditions. Here, the concept is used to penalize the term $\sigma(x, y, u) = -u^\top g(x, y)$ and, thus, gaining stronger conditions. Interesting for the

5.1 Properties in the General Case

linear case is that they also showed that a problem with a convex upper level function G and affine linear functions f and g already has a partially calm function σ.

Chen and Florian [23] showed that the Arrow-Hurwicz-Uzawa constraint qualification (see Arrow et al. [3]) does not hold for the KKT-reformulation of a quadratic bilevel problem, accessorily to the fact that also (MFCQ) is not valid. The latter was also shown by Ye et al. [98].

Ye and Zhu combined in their recent work [97] both, the KKT and the optimal value function approach, in one model and derived new and weaker constraint qualifications (called weakly calm) for which S- and B-stationary points exists. But, their motivation here was clearly nonconvex models for which neither the KKT approach nor the optimal value function approach itself fulfill the above mentioned optimality conditions.

Dempe and Dutta [31] employed a linear bilevel example in order to show that at global solutions the lower level problem may violate the (LICQ) (because it is degenerated), although the joint feasible set M is not degenerated. Along with an quadratic example – that shows that only for some x, the general lower level problem might violate the Slater condition and, thus, is not equal to its KKT reformulation – and some other observations, they show that bilevel problems are not a special case of MPCCs.

Another specialty of this approach is the observation that the KKT reformulation of the bilevel problem only coincides with its original in the sense of global optima. Even if the problem is convex and a usual regularity assumption is fulfilled, local optima may differ, which was also shown by Dempe and Dutta [31]. This has very strong impact on the algorithms based on the KKT approach since local optimal solutions found for reformulation (5.1) may not be locally optimal for the bilevel problems. In addition, error estmistation is fairly difficult due to the special structure.

5. KARUSH-KUHN-TUCKER APPROACH

5.2 Properties in the Linear Case

The linear bilevel problem (3.1) does not need any additional regularity assumptions due to the inherent structure and its (deterministic) KKT reformulation looks like

$$\min_{x,y,u,\mu} \quad c^\top x + d^\top y \tag{5.2}$$

$$\text{s.t. } Ax = b \tag{5.3}$$

$$Wy = h - Tx \tag{5.4}$$

$$W^\top u + \mu = q \tag{5.5}$$

$$\mu^\top y = 0 \tag{5.6}$$

$$x, y, u, \mu \geq 0 \tag{5.7}$$

where μ is a slack varibale that is introduced for readability.

So, as stated in the foregoing section, the linear KKT reformulation does not fulfill the MFCQ, LICQ, and Slater CQ.

Fortuny-Amat and McCarl [42] rewrote the complementarity slackness condition (5.6) using the big-M method under the assumption that the problem is bounded. Therefore, a binary variable z is introduced and the term is replaced by the two constraints $y \leq z M$ and $\mu \leq (1 - z) M$, where M is a large positive constant. This transformation is also called Fortuny-Amat transformation and results in a (larger) mixed-integer programming problem that can be optimized using a MILP solver. But, the large size of the augmented problem may produce high computation times when it comes to bigger problem sizes. Additionally, Gabriel and Leuthold [44] showed that the selection of a particular parameter M is often troublesome and that a solution can be extremely sensitive to its value. Bialas and Karwan [12] mention that if M is chosen large enough, some integer programming solutions may reduce to an enumeration scheme of all combinations $\mu = 0$ and $y = 0$. Nonetheless, this approach is widely used in the field of applications due to its simplicity and the very well developed MILP solvers.

But, also different approaches have been made. For example, Bard and Falk [8] transformed the complementarity constraint $\mu^\top y = 0$ into the terms

$$\sum_i [\min(0, \lambda_i) + \mu_i] = 0 \quad \text{and} \quad \lambda_i - y_i + \mu_i = 0, i = 1, \ldots, m$$

5.2 Properties in the Linear Case

and then used a branch-and-bound technique proposed by Falk and Soland [39] in order to find a global optimum. However, limited computational experience is reported.

Bialas et al. [13] presented the parametric complementary pivot (PCP) algorithm which adds the constraint $q^\top y \geq \alpha$ to the problem (5.2) – (5.4), (5.6), as well as (5.7) and changes (5.5) into $-\varepsilon Bx + W^\top u + \mu = q$ where ε is a suitably small positive scalar and B can be any negative-definite matrix. This leads to a small perturbation of the original KKT problem. The value of α is reasonably increased in every step until no feasible solution exists. The algorithm can be viewed as an implicit enumeration scheme of the lower-level optimal bases. Assumptions are that the problem is not degenerated and the lower level does not have multiple optima.

Hu et al. [50] proposed an algorithm for linear programs with linear complementarity constraints (with the KKT reformulation of bilevel problems in mind). It belongs to the class of enumerative algorithms which deploys a heuristic to branchen on the solution faster. It will be extensively exploited in the next chapter.

Rather recently, Siddiqui and Gabriel [83] derived a technique that reformulates the complementarity conditions using SOS1-type variables and Schur's Decomposition. They reformulate the complementarity term into

$$\begin{aligned} u - (v^+ - v^-) &= 0, \\ u &= \tfrac{u+\mu}{2}, \\ v^+ - v^- &= \tfrac{u-\mu}{2}, \\ v^\pm &\geq 0. \end{aligned}$$

In order to force each vector v_i^+, v_i^- to be as close to zero as possible, a penalty term $L(v^+ + v^-)$ is added to the objective. Both problems – with and without penalty term – are solved with changing penalty parameter L until $v^{+,\top} v^- = 0$ and the objective function value stays the same. The authors developed their algorithm for large-scale nonlinear MPEC models, but, also adress the case when all but the complementarity constraints are linear. Their motivation for developing this algorithm was a large-scale MPEC that refelcted the U.S. natural gas market and, therefore, they also included the case when the problem consists of finitely many stochastic scenarios in the follower's problem.

Introduction of Stochasticity

As stated in section 3.2, there are different approaches to establish stochastics in the problems. For computational reasons, the scenario set Ω will be discrete and finite

5. KARUSH-KUHN-TUCKER APPROACH

throughout, i.e., $\Omega = \{1, \ldots, N\}$. The uncertainty will be in the lower level only, which makes sense because the leader is most probably acquainted with its set of problems, but might lack in knowledge of the follower's decision making. Additionally, the nescience of the leader will only affect the right-hand side vector h and the objective function vector q of the follower.

The easiest approach is using the expected value function. It results in a risk-neutral stochastic program and can be represented as

$$\min_{x,y} \quad c^\top x + \sum_{\omega=1}^{N} \pi(\omega) d(\omega)^\top y(\omega) \qquad (5.8)$$

$$\text{s.t.} \quad Ax = b, x \geq 0 \qquad (5.9)$$

$$Wy(\omega) = h(\omega) - Tx, \qquad \forall \omega = 1, \ldots, N \qquad (5.10)$$

$$y(\omega) \geq 0, \qquad \forall \omega = 1, \ldots, N \qquad (5.11)$$

$$W^\top u(\omega) + \mu(\omega) = q(\omega), \qquad \forall \omega = 1, \ldots, N \qquad (5.12)$$

$$\mu(\omega) \geq 0, \qquad \forall \omega = 1, \ldots, N \qquad (5.13)$$

$$\mu^\top(\omega) y(\omega) = 0, \qquad \forall \omega = 1, \ldots, N \qquad (5.14)$$

with $0 \leq \pi(\omega) \leq 1$ being the weight of each scenario ω and $\sum_{\omega=1}^{N} \pi(\omega) = 1$. The follower's decision vector y is dependent on the scenario ω since the follower will choose differently dependent on the outcome of the stochastic event. Therefore, also the dual variables are dependent on ω. The leader has to anticipate any of those outcomes, but can incorporate the likeliness of the events. That is the reason for the different weights of the scenarios summing up to one.

Of course, non-anticipativity is assumed which ensures that x is not directly dependent on the stochastic outcome.

Few algorithms were developed that could solve this problem. The problem is that, due to the stochasticity, the problem size might increase very fast if the number of scenarios increases. And algorithms that were developed for small-scale problems might not automatically terminate for larger ones in reasonable time.

As above mentioned, Siddiqui and Gabriel [83] developed an algorithm that could also solve this problem through reformulating the complementarity constraints (5.14).

5.2 Properties in the Linear Case

Although, the authors do not show whether the algorithm solves the problem globally or only locally.

In addition, the algorithms for linear stochastic MPCCs mentioned at the end of section 1.4 could be used to solve the above problem.

5. KARUSH-KUHN-TUCKER APPROACH

6

Stolibi – An Algorithm for Linear Stochastic Bilevel Problems

As presented in the previous chapter, the KKT conditions produce a quadratic model due to the complementarity constraints. Those constraints are often replaced by constraints containing binary variables and a big-M scalar and then optimized using mixed-integer linear programming solvers such as IBM ILOG CPLEX Optimization Studio [53], Gurobi Software [47], or SCIP [100]. But, there are two drawbacks of such an approach. Normally, it is limited to problems with a bounded feasible region only, otherwise no such big-M would exist or, if computed with some constant M, the solution of the "big-M" problem would not be correct. On the other hand, the computation of the accurate size of the constant M is not trivial. If M is too large, it can lead to substantial round-off errors yielding an incorrect optimal solution; and if it is too small, the solution is not correct.

Below, an algorithm whose deterministic counterpart stems from Hu et al. [50] will be proposed. It was developed to globally solve linear problems with complementarity constraints (LPCC) having linear bilevel problems in mind. Still, Hu et al. [50] did not analyze or take advantage of the special structure that linear bilevel problems exhibit. A follow-up paper by Hu et al. [51] shows applications and how other problem classes can be reformulated as LPCCs as well as computational experiences with their algorithm.

The code presented in section 6.5 was implemented as a research code and is named stolibi.

6. STOLIBI – AN ALGORITHM FOR LINEAR STOCHASTIC BILEVEL PROBLEMS

Hu et al. [50] start from the big-M formulation and develop it into a parameter-free integer-programming-based cutting-plane algorithm. The method terminates with one out of three mutually exclusive conclusions:

- the KKT problem is infeasible,

- the KKT problem is feasible, but unbounded, or

- the KKT problem is feasible and attains a finite optimal solution.

Apart from the preprocessing and some special recovery procedure, the algorithm works on subproblems of the dual of the "big-M" model and an integer set Z "in the spirit of Benders' decomposition" (see Hu et al. [50]). The set is initialized as $Z = \{0,1\}^m$ and collects all satisfiability cuts that are generated during the procedure until the set is empty or a certificate of unboundedness is found. The branching procedure is aided by valid upper bounds on the (dual) optimal value.

The algorithm will be specialized to the case of linear stochastic bilevel problems and a decomposition method will be embeded.

Input data and results will be provided and the chapter will close with an evaluation of the refined algorithm.

6.1 Preliminaries

Hu et al. [50] start with a linear problem with complementarity constraints (LPCC) of the form

$$\min \quad c^\top x + d^\top y$$
$$\text{s.t.} \quad Ax + By \geq f \quad (6.1)$$
$$0 \leq y \perp q + Nx + My \geq 0,$$

where $a \perp b$ means that the two vectors are orthogonal, i.e., $a^\top b = 0$. It is $c \in \mathbb{R}^n, d, q \in \mathbb{R}^m, f \in \mathbb{R}^k, A \in \mathbb{R}^{k \times n}, B \in \mathbb{R}^{k \times m}, M \in \mathbb{R}^{m \times m}$ and $N \in \mathbb{R}^{m \times n}$.

Hu et al. [50] then state that "the LPCC (6.1) is equivalent to the minimization of a large number of linear programs, each defined on one piece of the feasible region of the LPCC", namely for each subset $\alpha \subset \{1, \ldots, m\}$ with complement $\bar{\alpha}$ the problems

6.1 Preliminaries

$LP(\alpha)$:

$$\min \quad c^\top x + d^\top y$$
$$\text{s.t.} \quad Ax + By \geq f$$
$$(q + Nx + My)_\alpha \geq 0 = y_\alpha, \tag{6.2}$$
$$(q + Nx + My)_{\bar{\alpha}} = 0 \leq y_{\bar{\alpha}}$$

The correspondence is of the following types (see page 447 in Hu et al. [50]):

- The LPCC (6.1) is infeasible if and only if the $LP(\alpha)$ is infeasible for all $\alpha \subset \{1, \ldots, m\}$;

- The LPCC (6.1) is feasible, but unbounded if and only if there exist some $\alpha \subset \{1, \ldots, m\}$ for which $LP(\alpha)$ is feasible and has an unbounded objective;

- The LPCC (6.1) is feasible and has a finite optimal value if and only if there exist some $\alpha \subset \{1, \ldots, m\}$ for which $LP(\alpha)$ is feasible and every such feasible $LP(\alpha)$ has a finite optimal objective value. In this case, the optimal objective value of the LPCC (6.1) (denoted by $LPCC_{min}$) is the minimum of the optimal objective values of all such feasible $LP(\alpha)$.

As stated in the previous chapter, the reformulated linear stochastic and risk-neutral bilevel problem using the KKT approach is of the form

$$\min \left\{ c^\top x + \sum_{\omega=1}^{N} \pi_\omega \cdot d_\omega^\top y(\omega) \;:\; \begin{array}{ll} Ax = b, \, x \geq 0 & \\ Wy(\omega) + Hx = h_\omega, & \forall \omega = 1, \ldots, N \\ W^\top u(\omega) \leq q_\omega, & \forall \omega = 1, \ldots, N \\ y(\omega) \geq 0, & \forall \omega = 1, \ldots, N \\ y(\omega)^\top (q_\omega - W^\top u(\omega)) = 0, & \forall \omega = 1, \ldots, N \end{array} \right\}$$

6. STOLIBI – AN ALGORITHM FOR LINEAR STOCHASTIC BILEVEL PROBLEMS

or with slack variable $\mu(\omega)$

$$\min \left\{ c^\top x + \sum_{\omega=1}^{N} \pi_\omega \cdot d_\omega^\top y(\omega) \;:\; \begin{array}{ll} Ax = b,\, x \geq 0 & \\ Wy(\omega) + Hx = h_\omega, & \forall \omega = 1,\ldots,N \\ W^\top u(\omega) + \mu(\omega) = q_\omega, & \forall \omega = 1,\ldots,N \\ \mu(\omega), y(\omega) \geq 0, & \forall \omega = 1,\ldots,N \\ y(\omega)^\top \mu(\omega) = 0, & \forall \omega = 1,\ldots,N \end{array} \right\} \quad (6.3)$$

and it can be translated into LPCC (6.1) form via

$$x^* = (x, y(1), \ldots, y(N))^\top, \; y^* = (u(1), \ldots, u(N), \mu(1), \ldots, \mu(N))^\top,$$

$$c^* = (c, \pi_1 \cdot d(1), \ldots, \pi_N \cdot d(N))^\top, \qquad d^* = 0,$$

$$A^* = \begin{pmatrix} A & 0 & \cdots & 0 \\ H & W & & \\ \vdots & & \ddots & \\ H & & & W \\ 0 & & & \\ & \ddots & & \\ & & & 0 \end{pmatrix}, \; B^* = \begin{pmatrix} 0 & & \cdots & & 0 \\ 0 & & & & \\ & & \ddots & & \\ W^\top & & & & 0 \\ & & \ddots & & \\ & & & & \\ & W^\top & & I_{n_2} & \\ & & \ddots & & \\ & & & W^\top & I_{n_2} \end{pmatrix}, \; f^* = \begin{pmatrix} b \\ h_1 \\ \vdots \\ h_N \\ q_1 \\ \vdots \\ q_N \end{pmatrix}$$

$$\text{and } N^* = \begin{pmatrix} 0 & 0 \\ 0 & I_{N \cdot n_2} \end{pmatrix}, \qquad q^* \equiv 0, \qquad M^* \equiv 0,$$

where I_n is the n–dimensional identity matrix. For readability, it is assumed that the first constraint of the LPCC is an equality constraint, i.e., $Ax + By = f$. Otherwise, the above matrices A^* and B^* as well as the vector f^* had to be of double size where each row occurs twice, once with a positive sign and once with a negative.

Here, it becomes a bit clearer that the already quite complex deterministic linear bilevel problem grows even bigger when passing to its expected-value-based stochastic version with a finite set of scenarios Ω. When a different risk functional is used that does not have a linear programming description and cannot be transposed into one, this equivalence is lost. This is also the case when Ω is a continuous set leading to a nonlinear term in the objective function and infinitely many constraints.

Now, following the standard approach would be to introduce a binary vector $z \in \{0,1\}^m$ and a positive scalar θ, replace the complementarity condition

$$y^\top (q + Nx + My) = 0$$

6.1 Preliminaries

using the Fortuny-Amat transformation, ending up with the mixed-binary model

$$
\begin{aligned}
\min \quad & c^\top x + d^\top y \\
\text{s.t.} \quad & Ax + By \geq f \\
& \theta z \geq q + Nx + My \geq 0, \\
& \theta(1-z) \geq y \geq 0, \\
& z \in \{0,1\}^m
\end{aligned} \quad (6.4)
$$

and optimizing it with a standard MILP solver (if θ is known to be of appropriate size).

But, following Hu et al. [50], the big-M formulation will just be conceptual as a means to further analyze the problem. To do so, the authors assumed θ to be given and z to be a parameter. Thus, the problem

$$
\begin{aligned}
\min \quad & c^\top x + d^\top y \\
\text{s.t.} \quad & Ax + By \geq f && (\lambda), \\
& Nx + My \geq -q && (u^-), \\
& -Nx - My \geq q - \theta z && (u^+), \\
& -y \geq -\theta(1-z) && (v), \\
& y \geq 0
\end{aligned} \quad (6.5)
$$

denoted by $\mathrm{LP}(\theta, z)$ is fully linear and can be dualized (the dual variables were given in the parentheses in the above model):

$$
\begin{aligned}
\max \quad & f^\top \lambda + q^\top(u^+ - u^-) - \theta[z^\top u^+ + (1-z)^\top v] \\
\text{s.t.} \quad & A^\top \lambda - N^\top(u^+ - u^-) = c, \\
& B^\top \lambda - M^\top(u^+ - u^-) - v \leq d, \\
& \lambda, u^+, u^-, v \geq 0
\end{aligned} \quad (6.6)
$$

The dual parametric problem is named $\mathrm{DLP}(\theta, z)$. The next proposition is crucial.

Proposition 6.1 (Hu et al. [50])
The following three statements hold:

6. STOLIBI – AN ALGORITHM FOR LINEAR STOCHASTIC BILEVEL PROBLEMS

a) Any feasible solution (x^0, y^0) of the LPCC (6.1) induces a pair (θ_0, z^0), where $\theta_0 > 0$ and $z^0 \in \{0,1\}^m$, such that the triple (x^0, y^0, z^0) is feasible to the mixed-binary model (6.4) for all $\theta \geq \theta_0$. Such z^0 has the property that

$$(q + Nx^0 + My^0)_i > 0 \Rightarrow z_i^0 = 1,$$
$$y_i^0 > 0 \Rightarrow z_i^0 = 0. \tag{6.7}$$

b) On the other hand, if (x^0, y^0, z^0) is feasible to the mixed-binary model (6.4) for some $\theta > 0$, then (x^0, y^0) is feasible to the LPCC (6.1).

c) If (x^0, y^0) is an optimal solution of the LPCC (6.1), then it is optimal to problem $LP(\theta, z^0)$ (6.5) for all $\theta \geq \theta_0$ (where θ_0 is the in a) induced scalar) and every z^0 that satisfies (6.7). Moreover, for each $\theta > \theta_0$, any optimal solution $(\hat{\lambda}, \hat{u}^+, \hat{u}^-, \hat{v})$ of the $DLP(\theta, z^0)$ (6.6) satisfies

$$z^{0\top} \hat{u}^+ + (1 - z^0)^\top \hat{v} = 0 \tag{6.8}$$

Especially c) represents an interesting fact that will be very useful. It shows that for every optimal solution (x^0, y^0) of the LPCC (6.1), there exists at least one binary vector z^0 such that the term in the dual problem's objective function

$$\theta[z^{0\top} u^+ + (1 - z^0)^\top v]$$

can be dropped provided that the following implications hold

$$z_i^0 = 1 \quad \Rightarrow \quad u_i^+ = 0 \quad \text{and} \quad z_i^0 = 0 \quad \Rightarrow \quad v_i = 0.$$

Corollary 6.2
Any (x^0, y^0) that is optimal for the LPCC (6.1) induces a pair (θ^0, z^0), as described in Proposition 6.1 a), such that the objective function value of $LP(\theta^0, z^0)$ (6.5) is minimal among the objective function values of all problems $LP(\theta^0, z), z \in \{0,1\}^m$.
Moreover, the same binary vector z^0 indicates the problem $DLP(\theta^0, z^0)$ (6.6) which has the lowest objective function value among all problems $DLP(\theta^0, z)$ (6.6) $z \in \{0,1\}^m$.

Proof:

The first assertion is due to the fact that

- both problems, LPCC (6.1) and $LP(\theta^0, z), z \in \{0,1\}^m$, have the same objective function,

- the feasibility set of $\mathrm{LP}(\theta^0, z)$, $z \in \{0,1\}^m$ is only a subset of the feasibility set of LPCC (6.1),

and Proposition 6.1 c).
The second assertion is due to the duality of $\mathrm{DLP}(\theta^0, z)$ (6.6) and $\mathrm{LP}(\theta^0, z)$. □

6.2 Decomposition

In the stochastic case, the dual $\mathrm{DLP}(\theta, z)$ is of the form

$$\max \left\{ \begin{array}{llll} \sum_{\omega=1}^{N} q_\omega^\top \lambda_A^\omega & -\theta \cdot \left(\sum_{\omega=1}^{N} z_\omega^\top v_1^\omega + \sum_{\omega=1}^{N} (1-z_\omega)^\top v_2^\omega \right) + b^\top \lambda_B^0 + \sum_{\omega=1}^{N} h_\omega^\top \lambda_B^\omega : & & \\ W \lambda_A^\omega & & = 0 & \forall \omega \\ \lambda_A^\omega & -v_1^\omega & \leq 0 & \forall \omega \\ & v_1^\omega & \geq 0 & \forall \omega \\ & A^\top \lambda_B^0 + \sum_{\omega=1}^{N} H^\top \lambda_B^\omega & \leq c & \\ & -v_2^\omega \qquad +W^\top \lambda_B^\omega & \leq \pi_\omega d_\omega & \forall \omega \\ & v_2^\omega & \geq 0 & \forall \omega \end{array} \right\}$$

(6.9)

Please note that, due to the equality conditions in the primal stochastic problem, the variables $\lambda_A^\omega, \lambda_B^\omega$, and λ_B^0 are not limited to the nonnegative orthant, but real decision vectors.

It is remarkable in the stochastic case that if it was not for the term

$$\theta \cdot \left(\sum_{\omega=1}^{N} z_\omega^\top v_1^\omega + \sum_{\omega=1}^{N} (1-z_\omega)^\top v_2^\omega \right), \tag{6.10}$$

the problem would be decomposable into two or more subproblems. But, due to (6.8), in the feasible and finite case it holds that this term vanishes in the dual problem $\mathrm{DLP}(\theta, z^0)$ with optimal solution vector z^0. This necessary optimality conditions motivates to define two sets for the stochastic case

$$I_1(z_\omega) := \{ i \,:\, v_{1,i}^\omega = 0 \} = \{ i \,:\, z_{\omega,i} = 1 \}$$

6. STOLIBI – AN ALGORITHM FOR LINEAR STOCHASTIC BILEVEL PROBLEMS

and
$$I_2(z_\omega) := \{i : v_{2,i}^\omega = 0\} = \{i : z_{\omega,i} = 0\}$$
where $z = (z_1, \ldots, z_N)$ and each of the sets are defined for every component vector of z.

Then, the term (6.10) can be dropped from the dual objective function in $\mathrm{DLP}(\theta, z)$ and the problem can be decomposed into the following.

The **first subproblem** is of the form

$$\max \left\{ \sum q_\omega^\top \lambda_A^\omega \ : \ \begin{array}{ll} W\lambda_A^\omega = 0, & \forall \omega = 1, \ldots, N \\ \lambda_A^\omega - v_1^\omega \leq 0, & \forall \omega = 1, \ldots, N \\ v_{1,i}^\omega = 0, \, i \in I_1(z_\omega), & \forall \omega = 1, \ldots, N \\ v_1^\omega \geq 0, & \forall \omega = 1, \ldots, N \end{array} \right\} \quad (6.11)$$

The constraint $\lambda_A^\omega - v_1^\omega \leq 0$ can be redefined since v_1 only presents a slack variable here. If $i \in I_1(z_\omega)$, then $v_{1,i} = 0$ and the constraint becomes $\lambda_{A,i}^\omega \leq 0$. For $i \notin I_1(z_\omega)$, $v_{1,i}$ can be chosen to be any positive value and the constraint becomes dispensable since v_1 is also no part of the objective function. Thus, the constraint can be rewritten as $\lambda_{A,i}^\omega \leq 0$ for $i \in I_1(z_\omega)$ and the remaining constraints containing v_1 can be dropped.

Moreover, subproblem (6.11) actually consists of N subproblems because the decision vectors λ_A^ω are not coupled. So, the first subproblem (6.11) decomposes into N problems

$$S(z_\omega) := \max \left\{ q_\omega^\top \lambda \ : \ W\lambda = 0, \ \lambda_i \leq 0, \ i \in I_1^\omega(z_\omega) \right\} \in \{0, \infty\} \quad (6.12)$$

These problems are homogeneous, i.e., $\lambda = 0$ is always a feasible solution with objective value 0. As soon as there exists a $\hat{\lambda}$ with $W\hat{\lambda} = 0$, but $q_\omega^\top \hat{\lambda} > 0$, the problem is unbounded and $S(z_\omega) = \infty$.

The **second subproblem** of $\mathrm{DLP}(\theta, z)$ (6.9) also possesses a special structure

$$\psi(z) = \max \left\{ b^\top \lambda_B^0 + \sum_{\omega=1}^N h_\omega^\top \lambda_B^\omega \ : \ \begin{array}{l} A^\top \lambda_B^0 + \sum_{\omega=1}^N H^\top \lambda_B^\omega \leq c, \\ \lambda_B^\omega \in P(z_\omega), \quad \forall \omega = 1, \ldots N \end{array} \right\} \quad (6.13)$$

where
$$P(z_\omega) = \{\lambda_B^\omega \in \mathbb{R}^{m_2} \ : \ W^\top \lambda_B^\omega - v_2^\omega \leq \pi_\omega d_\omega, \ v_{2,i}^\omega = 0, \, i \in I_2(z_\omega), \, v_2^\omega \geq 0\}.$$

6.2 Decomposition

As done in the first subproblem, the constraint $W^\top \lambda_B^\omega - v_2^\omega \leq \pi_\omega d_\omega$ can be rewritten. For $i \in I_2(z_\omega)$, it holds $v_{2,i}^\omega = 0$ and, thus, $(W^\top \lambda_B^\omega)_i \leq \pi_\omega d_{\omega,i}$. On the other hand, if $i \notin I_2(z_\omega)$, $v_{2,i}^\omega$ can be chosen to be any positive (and $-v_{2,i}^\omega$ to be any negative) value making the constraint disposable. Summarizing, P possesses the description

$$P(z_\omega) = \{\lambda_B^\omega \in \mathbb{R}^{m_2} : (W^\top \lambda_B^\omega)_i \leq \pi_\omega d_{\omega,i}, \, i \in I_2(z_\omega)\}.$$

In addition, the structure of the second subproblem allows for the use of the Dantzig-Wolfe decomposition. Therefore, assume that the extreme points and extreme rays of the subproblems $P(z_\omega)$ are known:

$$P(z_\omega) = \text{conv}\{p_1^\omega, \ldots, p_t^\omega\} + \text{cone}\{r_1^\omega, \ldots, r_a^\omega\}$$

where p_i^ω are the extreme points and r_j^ω represent the extreme rays. The following equivalence holds

$$\lambda \in P(z_\omega) \quad \Leftrightarrow \quad \lambda = \sum_{i=1}^{t} \gamma_i^\omega p_i^\omega + \sum_{j=1}^{a} \phi_j^\omega r_j^\omega$$

where $\sum_{i=1}^{t} \gamma_i^\omega = 1$ and $\gamma_i^\omega, \phi_j^\omega \geq 0$.

The complete master problem reads

$$\max \left\{ \begin{array}{l} b^\top \lambda_B^0 + \sum_{\omega=1}^{N} \left(\sum_{i=1}^{t} \gamma_i^\omega p_i^\omega h_\omega^\top + \sum_{j=1}^{a} \phi_j^\omega r_j^\omega h_\omega^\top \right) : \\ A^\top \lambda_B^0 + \sum_{\omega=1}^{N} \left(\sum_{i=1}^{t} \gamma_i^\omega p_i^\omega H^\top + \sum_{j=1}^{a} \phi_j^\omega r_j^\omega H^\top \right) \leq c, \\ \phantom{A^\top \lambda_B^0 + \sum_{\omega=1}^{N} \left(\sum_{i=1}^{t} \gamma_i^\omega p_i^\omega H^\top \right.} \sum_{i=1}^{t} \gamma_i^\omega = 1, \quad \forall \omega \\ \phantom{A^\top \lambda_B^0 + \sum_{\omega=1}^{N} \left(\sum_{i=1}^{t} \gamma_i^\omega p_i^\omega H^\top \right.} \gamma_i^\omega, \phi_j^\omega \geq 0, \quad \forall \omega, \forall i, \forall j \end{array} \right\}$$

The extreme points and rays can be computed scenariowise using the pricing problems

$$\max\left\{ \left(h_\omega^\top - q_\omega H^\top\right) \lambda_B^\omega : (W^\top \lambda_B^\omega)_i \leq \pi_\omega d_{\omega,i}, \, i \in I_2(z_\omega)\right\}.$$

The dual form of the KKT reformulated bilevel problem (6.3) and the special property of Proposition 6.1 c) opened the opportunity to use different decomposition techniques for the stochastic linear bilevel as was presented in this section. This is a new approach in the field of stochastic bilevel programming to the author's knowledge which will provide the possibility of parallelization and, thus, to find solutions faster.

6. STOLIBI – AN ALGORITHM FOR LINEAR STOCHASTIC BILEVEL PROBLEMS

6.3 Coverage of all Cases

The subsequent investigation follow Hu et al. [50] with emphazise on those issues that are relevant in the stochastic case.

If the original KKT reformulated bilevel problem (6.3) is solvable to optimality, then, the dual problem $\mathrm{DLP}(\theta, z), z \in \{0,1\}^m$, with the smallest optimal solution value provides the binary vector z which also induces the primal problem's optimal solution, see Corollary 6.2. Hu et al. [50] provide methods how to check problem (6.3) for unboundedness and infeasibility that will be presented now.

The following problem is of interest

$$\mathbb{R} \cup \{\pm\infty\} \ni \varphi(z) = \max_{\lambda, u^{\pm}, v} \quad f^\top \lambda + q^\top(u^+ - u^-)$$
$$\text{s.t.} \quad A^\top \lambda - N^\top(u^+ - u^-) = c,$$
$$B^\top \lambda - M^\top(u^+ - u^-) - v \leq d, \quad (6.14)$$
$$\lambda, u^+, u^-, v \geq 0$$
$$z^\top u^+ + (1-z)^\top v \leq 0$$

The above is proposed in section 2.1 of Hu et al. [50]. For linear stochastic bilevel programs, subproblems $\psi(z)$ and $S(z_\omega)$ (6.12), $\omega = 1, \ldots, N$ take the role of $\varphi(z)$[1] (6.13). As the former, problem (6.14) was motivated by Proposition 6.1 c).

In addition, the homogenization of $\varphi(z)$ (6.14) is

$$\{0, \infty\} \ni \varphi_0(z) = \max_{\lambda, u^{\pm}, v} \quad f^\top \lambda + q^\top(u^+ - u^-)$$
$$\text{s.t.} \quad A^\top \lambda - N^\top(u^+ - u^-) = 0,$$
$$B^\top \lambda - M^\top(u^+ - u^-) - v \leq 0, \quad (6.15)$$
$$\lambda, u^+, u^-, v \geq 0$$
$$z^\top u^+ + (1-z)^\top v \leq 0$$

In the stochastic case, this problem corresponds to subproblems $S(z_\omega)$ (6.12), $\omega = 1, \ldots, N$ and the homogenization of $\psi(z)$ (6.13) which will be named $\psi_0(z)$.

---

[1] Throughout this thesis, the reference to optimality problems either is made in the standard way by numbering or as above via reference to the optimal value.

6.3 Coverage of all Cases

It is known from linear algebra that for any pair (c, d) for which $\varphi(z)$ is feasible, it holds

$$\varphi(z) < \infty \quad \Leftrightarrow \quad \varphi_0(z) = 0.$$

If $\varphi(z)$ is unbounded, then of course, $\varphi_0(z)$ is unbounded, too. If $\varphi_0(z)$ is unbounded, the latter reverse implication only follows if $\varphi(z)$ is feasible; otherwise, $\varphi(z)$ could be infeasible.

The authors Hu et al. [50] additionally show (Proposition 2.2) that

$$\varphi_0(z) = 0 \quad \Leftrightarrow \quad LP(\alpha) \text{ (6.2) is feasible where } \alpha = supp(z) \qquad (6.16)$$

which also induces that $\varphi_0(z) = \infty$ is responsible for the infeasibility of the primal problems $LP(\alpha)$ (6.2). Especially, if $\varphi(z)$ is infeasible and $\varphi_0(z) = 0$, then it follows by (6.16) that the corresponding $LP(\alpha)$ is unbounded. Besides, it holds by Farkas' Lemma that $LP(\alpha)$ is infeasible if $\varphi(z)$ is infeasible and $\varphi_0(z) = \infty$ for $\alpha = supp(z)$.

The key set, which will be of interest in the algorithm, can now be defined as

$$Z = \{z \in \{0, 1\}^m : \varphi_0(z) = 0\}.$$

It contains all those z for which the primal problems $LP(\alpha)$ are feasible (unbounded or bounded) for $\alpha = supp(z)$. If this set is empty, it is known from section 6.1 that the LPCC (6.1) is infeasible.

Thus, the new optimization problem will be

$$\min \varphi(z) \quad \text{s.t.} \quad z \in Z \qquad (6.17)$$

and it holds that

Theorem 6.3 (Hu et al. [50], Theorem 2.4)

a) LPCC (6.1) is infeasible if and only if $Z = \emptyset$ (i.e., $\min_{z \in Z} \varphi(z) = \infty$).

b) LPCC (6.1) is feasible, but unbounded, if and only if $\min_{z \in Z} \varphi(z) = -\infty$.

c) LPCC (6.1) is feasible and bounded if and only if $-\infty < \min_{z \in Z} \varphi(z) < \infty$.

In all cases coincide the optimal values $LPCC_{min} = \min_{z \in Z} \varphi(z)$.

Moreover, it holds for any $z \in \{0, 1\}^m$ for which $\varphi(z)$ is feasible that

$$LPCC_{min} \leq \varphi(z).$$

□

6. STOLIBI – AN ALGORITHM FOR LINEAR STOCHASTIC BILEVEL PROBLEMS

In the stochastic case, it follows that

$$Z^* = \{z = (z_1, \ldots, z_N) \in \{0,1\}^{m \times N} : S(z_\omega) = 0, \forall \omega \wedge \psi_0(z) = 0\}$$

where $\psi_0(z)$ is the homogenization of $\psi(z)$ (6.13) and the problem is to minimize the function $\psi(z)$ over this set Z^*. As above, the following cases arise:

a) The KKT reformulated risk-neutral stochastic bilevel problem (6.3) is infeasible if and only if $Z^* = \emptyset$.

b) Problem (6.3) is feasible, but unbounded, if and only if $\min_{z \in Z^*} \psi(z) = -\infty$.

c) Problem (6.3) is feasible and bounded if and only if $-\infty < \min_{z \in Z^*} \psi(z) < \infty$.

In b) and c), $S(z_\omega)$ is not part of the objective function value since its value is zero due to the constraint $z \in Z^*$.

6.4 Search on the Binary Set

Throughout the algorithm of Hu et al. [50], different cuts will be added to the set $\{0,1\}^m$ in order to find a certificate of optimality, infeasibility, or unboundedness. This section provides the proofs that the generated cuts lead to the correct certificate.

Because of Proposition 6.1, the relationship between the following two feasibility sets is clear for the optimal, bounded solution, but not for the other cases.

$$\Xi = \left\{ \begin{array}{l} (\lambda, u^\pm, v) : \ A^\top \lambda - N^\top(u^+ - u^-) = c, \\ \qquad\qquad\qquad B^\top \lambda - M^\top(u^+ - u^-) - v \leq d, \\ \qquad\qquad\qquad \lambda, u^\pm, v \geq 0 \end{array} \right\}$$

is the feasibility set of problem $\mathrm{DLP}(\theta, z)$ (6.6) and

$$T(z) = \left\{ \begin{array}{l} (\lambda, u^\pm, v) : \ A^\top \lambda - N^\top(u^+ - u^-) = c, \\ \qquad\qquad\qquad B^\top \lambda - M^\top(u^+ - u^-) - v \leq d, \\ \qquad\qquad\qquad \lambda, u^\pm, v \geq 0 \\ \qquad\qquad\qquad z^\top u^+ + (1-z)^\top v \leq 0 \end{array} \right\}$$

represents the feasibility set of $\varphi(z)$ (6.14).

6.4 Search on the Binary Set

The following proposition will express the equivalence between those two sets. It is also stated in the paper Hu et al. [50] as Proposition 2.3 and its proof will be repeated here a bit more detailed.

Proposition 6.4
The subsequent holds for any $z \in \{0,1\}^m$:

1. An extreme point or extreme ray of $T(z)$ is also extreme in Ξ.

2. An extreme point or extreme ray of Ξ is also extreme in $T(z)$ if it is feasible.

Proof:
The assertions will only be shown for extreme points. Those for the rays can be shown equivalently.

To prove the first statement, assume that $(\lambda^p, u^{\pm,p}, v^p)$ is an extreme point of $T(z)$ (for any z). Then, of course, it is feasible to Ξ since $T(z) \subseteq \Xi$ for all z. To show that it is extreme, assume that there exist two other vectors in Ξ with

$$\begin{pmatrix} \lambda^p \\ u^{+,p} \\ u^{-,p} \\ v^p \end{pmatrix} = \alpha \begin{pmatrix} \lambda^1 \\ u^{+,1} \\ u^{-,1} \\ v^1 \end{pmatrix} + (1-\alpha) \begin{pmatrix} \lambda^2 \\ u^{+,2} \\ u^{-,2} \\ v^2 \end{pmatrix}$$

where $\alpha \in (0,1)$ is some scalar and $T(z) \not\ni (\lambda^i, u^{\pm,i}, v^i) \in \Xi$, $i = 1,2$. Then, it holds

$$0 \geq z^\top u^{+,p} + (1-z)^\top v^p = z^\top (\alpha u^{+,1} + (1-\alpha) u^{+,2}) + (1-z)^\top (\alpha v^1 + (1-\alpha) v^2).$$

Since all participating vectors and scalars on the right are nonnegative, it holds

$$z^\top \alpha u^{+,i} + (1-z)^\top \alpha v^i \leq 0 \quad \Rightarrow \quad z^\top u^{+,i} + (1-z)^\top v^i \leq 0$$

which includes $(\lambda^i, u^{\pm,i}, v^i) \in T(z)$, a contradiction.

The second assertion holds trivially due to $T(z) \subseteq \Xi$. $\square$

Therefore, assume that all extreme points and rays of Ξ are known, i.e., the vectors $\{(\lambda^{p_i}, u^{\pm,p_i}, v^{p_i})\}_{i=1}^K$ and $\{(\lambda^{r_j}, u^{\pm,r_j}, v^{r_j})\}_{j=1}^L$ define the K extreme points and L extreme rays of Ξ, respectively. Having that, the problem (6.18), $\min \varphi(z), z \in Z$, can

6. STOLIBI – AN ALGORITHM FOR LINEAR STOCHASTIC BILEVEL PROBLEMS

be rewritten as

$$\min_{z \in Z} \left[\begin{array}{l} \max_{\beta^p, \beta^r} \sum_{i=1}^{K} \beta^{p,i} \left(f^\top \lambda^{p_i} + q^\top (u^{+,p_i} - u^{-,p_i}) \right) + \sum_{j=1}^{L} \beta^{r,j} \left(f^\top \lambda^{r_j} + q^\top (u^{+,r_j} - u^{-,r_j}) \right) \\ \text{s.t.} \quad \sum_{i=1}^{K} \beta^{p,i} \left(z^\top u^{+,p_i} + (1-z)^\top v^{p_i} \right) + \sum_{j=1}^{L} \beta^{r,j} \left(z^\top u^{+,r_j} + (1-z)^\top v^{r_j} \right) \leq 0 \\ \sum_{i=1}^{K} \beta^{p,i} = 1 \\ \beta^p, \beta^r \geq 0 \end{array} \right] $$

(6.18)

as done in section 2.2 with the Dantzig-Wolfe Decomposition.

Two sets can be defined that are of importance here:

$$\mathcal{L} = \{ j \in \{1, \ldots, L\} : f^\top \lambda^{r_j} + q^\top (u^{+,r_j} - u^{-,r_j}) > 0 \}$$

which contains all indices of extreme rays with positive objective function value in the above problem; and

$$\mathcal{K} = \{ i \in \{1, \ldots, K\} : f^\top \lambda^{p_i} + q^\top (u^{+,p_i} - u^{-,p_i}) = \varphi(z) \text{ for some } z \in Z \}$$

which contains all those indices of extreme points that define the optimal solution for any $\varphi(z)$.

The set $Z = \{z \in \{0,1\}^m : \varphi_0(z) = 0\}$ can be redefined using the set $\mathcal{L}$:

Proposition 6.5 (Hu et al. [50], Proposition 3.1)

$$Z = \left\{ z \in \{0,1\}^m : \sum_{l: u_l^{+,r_j} > 0} z_l + \sum_{k: v_k^{r_j} > 0} (1-z)_k \geq 1 \ \forall j \in \mathcal{L} \right\}$$

□

The above representation of Z is due to the fact that problem $\varphi_0(z)$ (6.15) can also be defined using the extreme rays of Ξ. $\varphi_0(z)$ only has a bounded objective function value if the above inequalities describing Z hold true.

In addition to that, if a set $\mathcal{R} \subseteq \mathcal{L}$ is found such that

$$Z \subseteq \left\{ z \in \{0,1\}^m : \sum_{l: u_l^{+,r_j} > 0} z_l + \sum_{k: v_k^{r_j} > 0} (1-z)_k \geq 1 \ \forall j \in \mathcal{R} \right\} = \emptyset$$

6.4 Search on the Binary Set

then the LPCC (6.1) is infeasible, which holds due to Theorem 6.3 a) and the Proposition 6.5. Therefore, problem (6.18), $\min \varphi(z), z \in Z$, can again be rewritten as

$$\min_{z \in Z} \begin{bmatrix} \max_{\beta^p} \sum_{i=1}^{K} \beta^{p,i} \left(f^\top \lambda^{p_i} + q^\top (u^{+,p_i} - u^{-,p_i}) \right) \\ \text{s.t.} \sum_{i=1}^{K} \beta^{p,i} \left(z^\top u^{+,p_i} + (1-z)^\top v^{p_i} \right) \leq 0 \\ \sum_{i=1}^{K} \beta^{p,i} = 1 \\ \beta^p \geq 0 \end{bmatrix} \quad (6.19)$$

Because of the new definition of Z, it forces $\beta^{r,j}$ to be zero every time the corresponding extreme ray produces a positive objective function value, and the fact that if $f^\top \lambda^{r_j} + q^\top(u^{+,r_j} - u^{-,r_j}) < 0$, $\beta^{r,j} = 0$ would always be a feasible and optimal solution.

Now, the second set, $\mathcal{K}$, will be examined. Clearly, if $\mathcal{K} \neq \emptyset$, then $Z \neq \emptyset$ and LPCC (6.1) is feasible, but possibly unbounded. Due to Theorem 6.3 and the definition of $\mathcal{K}$, it holds that

$$\min_{i \in \mathcal{K}} f^\top \lambda^{p_i} + q^\top (u^{+,p_i} - u^{-,p_i}) \geq LPCC_{min}.$$

During the algorithm, different cuts will be generated for the binary set $\{0,1\}^m$. The following set descibes them

$$Z(P,R) = \left\{ \begin{array}{l} z \in \{0,1\}^m : \sum_{l:u_l^{+,r_j}>0} z_l + \sum_{k:v_k^{r_j}>0} (1-z)_k \geq 1 \ \forall j \in R \\ \sum_{l:u_l^{+,p_i}>0} z_l + \sum_{k:v_k^{p_i}>0} (1-z)_k \geq 1 \ \forall i \in P \end{array} \right\}$$

for $P \times R \subseteq \mathcal{K} \times \mathcal{L}$. The first line denotes ray cuts, whereas the second line describes point cuts.

Lateron, problem $\varphi(z)$ (6.14) will be minimized with respect to the subset $Z(P,R)$ of binaries with growing sets $P \subseteq \mathcal{K}$, collecting the extreme points, and $R \subseteq \mathcal{L}$, collecting extreme rays.

Proposition 6.6 (Hu et al. [50], Proposition 3.3)
If there exists $P \subseteq \mathcal{K}$ and $R \subseteq \mathcal{L}$ such that

$$\min_{i \in P} f^\top \lambda^{p_i} + q^\top (u^{+,p_i} - u^{-,p_i}) > LPCC_{min},$$

then $\operatorname{argmin}_{z \in Z} \varphi(z) \subseteq Z(P,R)$. □

6. STOLIBI – AN ALGORITHM FOR LINEAR STOCHASTIC BILEVEL PROBLEMS

From this it follows that if upon iterating the minimization problem $\min_{z \in Z} \varphi(z)$, optimality is not reached yet, the current set $Z(P, R)$ always contains the $\operatorname{argmin}_{z \in Z} \varphi(z)$. In other words, it cannot happen that optimal solutions get lost on the way. Additionally,

Proposition 6.7 (Hu et al. [50], Corollary 3.4)
If there exist $P \times R \subseteq \mathcal{K} \times \mathcal{L}$ with $P \neq \emptyset$ and $Z(P,R) = \emptyset$, then

$$LPCC_{min} = \min_{i \in P} f^\top \lambda^{p_i} + q^\top (u^{+,p_i} - u^{-,p_i}) \in \mathbb{R}$$

□

Altogether, it can be said that

Theorem 6.8 (Hu et al. [50], Theorem 3.5)

a) LPCC (6.1) is infeasible if and only if a subset $R \subseteq \mathcal{L}$ exists such that $Z(\emptyset, R) = \emptyset$;

b) LPCC (6.1) is feasible, but unbounded if and only if $Z(\mathcal{K}, \mathcal{L}) \neq \emptyset$;

c) LPCC (6.1) attains a finite value if and only if a pair exists $P \times R \subseteq \mathcal{K} \times \mathcal{L}$ with $P \neq \emptyset$ such that $Z(P, R) = \emptyset$.

This theorem is crucial for the algorithm, since the latter searches for sets $P \times R$ such that $Z(P, R) = \emptyset$. If in that case $P = \emptyset$, the LPCC is infeasible, and if $P \neq \emptyset$, the LPCC can be solved to optimality. The unboundedness can be found as presented in the previous section through $\varphi(z) = -\infty$ and $\varphi_0(z) = 0$.

According to the stochastic dual model (6.9) on page 67, a ray cut can be defined similarly

$$\sum_{\omega=1}^{N} \sum_{l:v_{1,l}^{\omega,*}>0} z_{\omega,l} + \sum_{\omega=1}^{N} \sum_{k:v_{2,k}^{\omega,*}>0} (1-z)_{\omega,k} \geq 1$$

if both subproblems, ψ (6.13) and at least one S-model (6.12), are unbounded. $v_1^{\omega,*}$ as well as $v_2^{\omega,*}$ are the slack variables of the extreme ray solution vectors (as analyzed in section 6.2). If only one of the subproblems is unbounded, the above cut simplifies to either

$$\sum_{\omega=1}^{N} \sum_{l:v_{1,l}^{\omega,*}>0} z_{\omega,l} \geq 1$$

6.4 Search on the Binary Set

if at least one S-model was unbounded, but ψ was bounded, or

$$\sum_{\omega=1}^{N} \sum_{k:v_{2,k}^{\omega,*}>0} (1-z)_{\omega,k} \geq 1$$

if ψ was unbounded, but all S-models were bounded.

For the point cut, the situation is simpler due to the fact that if all S-models are bounded, then, $S(z_\omega) = 0 \to \lambda_A^\omega = 0$ which also implies that $v_1^\omega = 0$. So, a point cut is of the form

$$\sum_{\omega=1}^{N} \sum_{k:v_{2,k}^{\omega,*}>0} (1-z)_{\omega,k} \geq 1$$

where $v_2^{\omega,*}$ are the slack variables of the optimal solution of ψ.

The values of the slack variables for the S-models can be easily found by:

$$\lambda_{A,l}^\omega > 0 \quad \Leftrightarrow \quad v_{1,l}^\omega > 0.$$

But, apart from the discussion in section 6.2, v_2^ω cannot be substituted from the model ψ in order to be able to define point and ray cuts. Remember that

$$P(z_\omega) = \{\lambda_B^\omega \in \mathbb{R}^{m_2} : (W^\top \lambda_B^\omega)_i \leq \pi_\omega d_{\omega,i}, i \in I_2(z_\omega)\}$$

but, only for $i \notin I_2(z_\omega)$, the vectors v_2^ω can be positive. Additionally, for different binary vectors z_ω, the sets $P(z_\omega)$ may differ a lot. This implies that an extreme ray or point that was found for one $P(z_\omega)$ cannot automatically be used for a different $P(z_\omega^*)$. Thus, the Dantzig-Wolfe decomposition discussed on the problem ψ remains theoretical for now. The subproblem ψ optimized in the next section is

$$\psi(z) = \max \left\{ \begin{array}{l} b^\top \lambda_B^0 + \sum_{\omega=1}^{N} h_\omega^\top \lambda_B^\omega : \\ A^\top \lambda_B^0 + \sum_{\omega=1}^{N} H^\top \lambda_B^\omega \leq c, \\ W^\top \lambda_B^\omega - v_2^\omega \leq \pi_\omega d_\omega, \quad \forall \omega = 1, \ldots, N \\ v_{2,i}^\omega = 0, \quad i \in I_2(z_\omega), \quad \forall \omega = 1, \ldots, N \\ v_2^\omega \geq 0, \quad \forall \omega = 1, \ldots, N \end{array} \right\}$$

The following section describes the procedure of the algorithm.

6. STOLIBI – AN ALGORITHM FOR LINEAR STOCHASTIC BILEVEL PROBLEMS

6.5 Pseudo-Code of Stolibi

The algorithm to be presented now consists of three parts: preprocessing, main procedure, and sparsification. While the first two steps are pretty much conventional, the third is not. Conceptually, sparsification is an optional heuristic to speed up branching in the main part. Otherwise, the latter could get very close to complete enumeration, in the worst case. In the following, each of the three parts is inspected closer. In addition, figures 6.1 -6.3 show the flowchart for each part.

6.5.1 Preprocessing

Preprocessing

Data Input: Two files [name1.lp name2.txt]
If error occurs, **STOP** with error description.
Else solve relaxed bilevel problem

$$\min\{c^\top x + \sum_{\omega=1}^{N} d_\omega^\top y(\omega) \,:\, \begin{array}{ll} Ax = b,\, x \geq 0, & \\ Wy(\omega) = h_\omega - Tx, & \forall \omega \\ y(\omega) \geq 0, & \forall \omega \\ W^\top u(\omega) + \mu(\omega) = q_\omega, & \forall \omega \\ \mu(\omega) \geq 0 & \forall \omega \end{array}\right\}$$

If this problem is infeasible, **STOP**
 – the original problem is infeasible.

If this is optimal and complementarity conditions are already fulfilled, **STOP**
 – the original problem is already solved to optimality.

Else construct binary set $Z = \mathrm{argmin}\{z : z \in \{0,1\}^{m*N}\}$ as well as dual models S and ψ, set upper bound P_{opt} to infinity and start with the main procedure.

The preprocessing starts with data handling. The algorithm's input comes in two files, the first of which (name1.lp) contains the upper level with its constraints $Ax = b, x \geq 0$ and objective function $c^\top x$ as well as the lower level constraints for one scenario:

$$\begin{aligned} \min \quad & c^\top x + d_1^\top y(1) \\ \text{s.t.} \quad & Ax = b,\, x \geq 0, \\ & Wy(1) = h_1 - Tx, \\ & y(1) \geq 0. \end{aligned}$$

6.5 Pseudo-Code of Stolibi

The second file (name2.txt) has to contain

1. the total number of scenarios (SCEN),

2. the dimension of the lower-level decision vector (SIZE),

3. a scalar parameter for the main procedure to estimate whether it pays to start sparsification (EPS),

4. the lower level cost data vectors for the individual scenarios (Q*),

5. the lower level right-hand side data vectors for all but the first scenarios (changeh*), and

6. the follower's upper level cost data vectors for the individual scenarios but the first (D*).

With the information of the two files, the relaxed bilevel problem

$$\min\{c^\top x + \sum_{\omega=1}^{N} d_\omega^\top y(\omega) \; : \; \begin{array}{ll} Ax = b\,,\; x \geq 0\,, & \\ Wy(\omega) = h_\omega - Tx\,, & \forall \omega \\ y(\omega) \geq 0\,, & \forall \omega \\ W^\top u(\omega) + \mu(\omega) = q_\omega\,, & \forall \omega \\ \mu(\omega) \geq 0 & \forall \omega \end{array} \right\} \quad (6.20)$$

is built and is then solved in the second part of the preprocessing. It is relaxed because it is missing the complementarity constraints $\mu(\omega)^\top y(\omega) = 0$, $\forall \omega = 1,\ldots N$.

Should this problem be infeasible, then, the original KKT reformulated bilevel problem is infeasible, too. In this case, the algorithm stops with this output message.

Additionally, if the optimal solution of the relaxed bilevel problem (6.20) already fulfills the complementarity constraints $\mu(\omega)^\top y(\omega) = 0$ for all $\omega = 1,\ldots N$, then the optimal solution for the original KKT reformulated bilevel problem is trivially found. If neither one is the case, the procedure goes on with constructing the binary set Z, which is realized through the usage of a minimization problem, initially on the total number of possibilities ($Z = \text{argmin}\{z : z \in \{0,1\}^{m*N}\}$). Next, the dual problems $S(z_\omega), \omega = 1,\ldots, N$ and the dual problem $\psi(z)$ are constructed. The upper bound, which will be updated every time a feasible and bounded solution is found, is set to infinity. A graphic representation of the preprocessing can be found in figure 6.1.

6. STOLIBI – AN ALGORITHM FOR LINEAR STOCHASTIC BILEVEL PROBLEMS

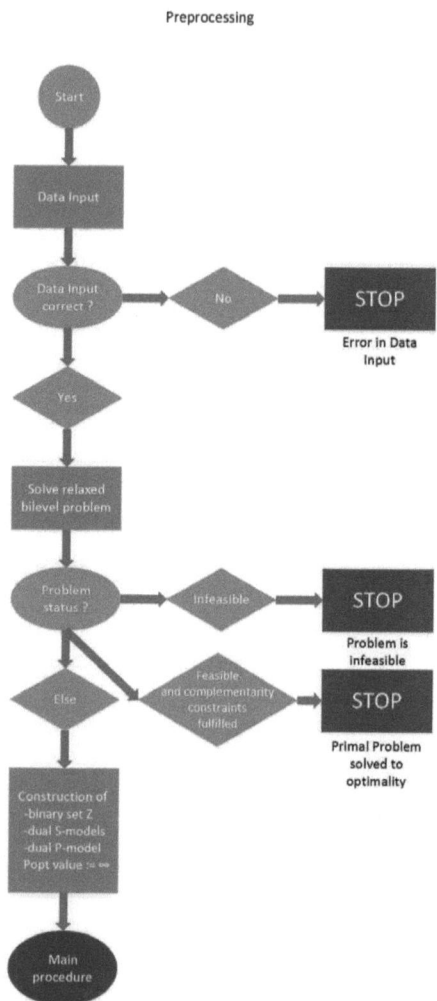

Figure 6.1: Flow chart of the preprocessing.

6.5.2 Main Procedure

For ease of representation, recall the following problems

$$\psi(z) = \max \left\{ \begin{array}{l} b^\top \lambda_B^0 + \sum_{\omega=1}^N h_\omega^\top \lambda_B^\omega : \\ A^\top \lambda_B^0 + \sum_{\omega=1}^N H^\top \lambda_B^\omega \leq c, \\ W^\top \lambda_B^\omega - v_2^\omega \leq \pi_\omega d_\omega, \quad \forall \omega = 1, \ldots, N \\ v_{2,i}^\omega = 0, \quad i \in I_2(z_\omega), \quad \forall \omega = 1, \ldots, N \\ v_2^\omega \geq 0, \quad \forall \omega = 1, \ldots, N \end{array} \right\}$$

and

$$S(z_\omega) := \max \left\{ q_\omega^\top \lambda : \quad W\lambda = 0, \quad \lambda_i \leq 0, \quad i \in I_1^\omega(z_\omega) \right\}.$$

Main Procedure

Data Input: $Z = \operatorname{argmin}\{z : z \in \{0,1\}^{m*N}\}$, ψ, and S-models

While $Z \neq \emptyset$
  Choose $z \in Z = \arg \min \{ \mathbb{1}^\top z : z \in \{0,1\}^{m*N}$, z fulfills all generated cuts $\}$.
  Solve $\psi(z)$
    If it is *unbounded*, solve all homogeneous models $S(z_\omega), \forall \omega = 1, \ldots, N$,
    add ray cut to Z, and start the sparsification procedure.
    If it is *infeasible*, check if $\psi_0(z)$ and all models $S(z_\omega)$ are bounded.
      If so, STOP – the primal bilevel problem is unbounded.
      If not, add ray cut to Z, and start the sparsification procedure.
    If it is *bounded*, solve all homogeneous models $S(z_\omega), \forall \omega = 1, \ldots, N$
      If at least one model is *unbounded*, find one extreme ray in every unbounded model, add the corresponding ray cut to Z, and start the sparsification procedure.
      If all S–models are *bounded*, add point cut to Z, and
        If $\psi(z) > P_{opt} + EPS$, start sparsification procedure,
        Else if $\psi(z) < P_{opt}$, update the bound P_{opt}, save the solution z_{opt},
        and start while-loop again,
          Else start while-loop again.
End
If no point cut has been added to Z ($P_{opt} = \infty$), **STOP** – the bilevel problem is infeasible.
Else STOP – optimal solution found with value P_{opt} and binary vector z_{opt}.

It is important to mention that after the sparsification procedure the algorithm starts over with the while-loop, but possibly with Z having more constraints or the

6. STOLIBI – AN ALGORITHM FOR LINEAR STOCHASTIC BILEVEL PROBLEMS

bound P_{opt} being tighter. More about this procedure in the next subsection.

So, the main procedure starts with the data that it gets from the preprocessing. At the first run, Z will have the total binary set $\{0,1\}^{m*N}$ as feasibility set because no cuts have been added to Z so far. Thus, the origin will be chosen to be the first z.

The authors of Hu et al. [50] did not specialize how to search on the set Z. I chose to start with binary vectors closer to the origin since these would restrict the dual set $\psi(z)$ the most, giving the possibility to lead to dual infeasibility (primal unboundedness) faster than the other way round. But this is just one way of doing it and it is not proven if one approach is faster than the other.

Having chosen a vector z, the algorithm goes on with solving $\psi(z)$. Of course, only three cases can be established for $\psi(z)$: unboundedness, infeasibility, or feasibility.

In the first case, also the S-models have to be checked for unboundedness in order to derive a feasible ray cut of the form

$$\sum_{\omega=1}^{N} \sum_{l:v_{1,l}^{\omega,*}>0} z_{\omega,l} + \sum_{\omega=1}^{N} \sum_{k:v_{2,k}^{\omega,*}>0} (1-z)_{\omega,k} \geq 1$$

– as described at the end of the previous section 6.4 – which is added to Z. Again, if only problem ψ is unbounded, but all S-models are bounded, the cut simplifies to only the right sum being greater or equal to one. The sparsification procedure starts with that cut then.

If the problem is infeasible, it has to be checked if a certificate of unboundedness for the primal problem is found. Therefore, $\psi_0(z)$ and all S-models will be checked. If at least one of the homogenization models is unbounded, the algorithm goes on with constructing a new ray cut for Z which will be built from the (extreme ray) solution of all unbounded homogeneous models. This means that the cut is of the same form as above

$$\sum_{\omega=1}^{N} \sum_{l:v_{1,l}^{\omega,*}>0} z_{\omega,l} + \sum_{\omega=1}^{N} \sum_{k:v_{2,k}^{\omega,*}>0} (1-z)_{\omega,k} \geq 1.$$

Again, the sparsification procedure will start with that cut.

But, if all models are bounded, the algorithm stops with the display `Primal problem is unbounded given that the homogenization is bounded. Stopped.` and the according binary z-vector.

6.5 Pseudo-Code of Stolibi

If $\psi(z)$ is bounded and feasible, it still has to be checked if all S-models are also bounded as functions of the same z. So, if at least one of these $S(z_\omega)$ is unbounded, a ray cut will be built:

$$\sum_{\omega=1}^{N} \sum_{l:v_{1,l}^{\omega,*}>0} z_{\omega,l} \geq 1.$$

Since only the S-problems are unbounded in that case, the cut is so short. The sparsification procedure takes that cut as starting point.

On the other hand, if all S-models are bounded, a feasible solution is found and a point cut can be added to Z as explained in section 6.4

$$\sum_{\omega=1}^{N} \sum_{k:v_{2,k}^{\omega,*}>0} (1-z)_{\omega,k} \geq 1.$$

Additionally in that case, it has to be checked if the optimal value is better than the current best solution value. If so, the algorithm updates the bound P_{opt} and saves the according binary vector as z_{opt}. The algorithm starts again with solving Z.

As proposed by Hu et al. [50], nothing more will be done if the optimal solution of $\psi(z)$ is only a "little bit" better than the current best bound. The authors argumented that in this case, the cut would not be promising enough for the sparsification procedure. In that case then, the algorithm start again with solving Z.

But, if the optimal solution value of $\psi(z)$ is greater than $P_{opt} + EPS$ (which defines the "little bit"), the cut seems promising for the sparsification procedure. The value of EPS can be chosen by the user, see the previous subsection.

The algorithm searches on the set Z until it is empty. In that case, the algorithm stops with either having found an optimal solution P_{opt} and the according binary vector z_{opt} or having found a certificate of infeasibility for the KKT reformulated bilevel problem.

Remark 1
If nonpositive extreme rays with positive objective value(s) exist in the above unbounded models, the dual problem is unbounded for all z. In such a case, the algorithm would add the cut

$$\sum_{i:z_i^*=0} z_i + \sum_{j:z_j^*=1} (1-z)_j \geq 1 \qquad (6.21)$$

6. STOLIBI – AN ALGORITHM FOR LINEAR STOCHASTIC BILEVEL PROBLEMS

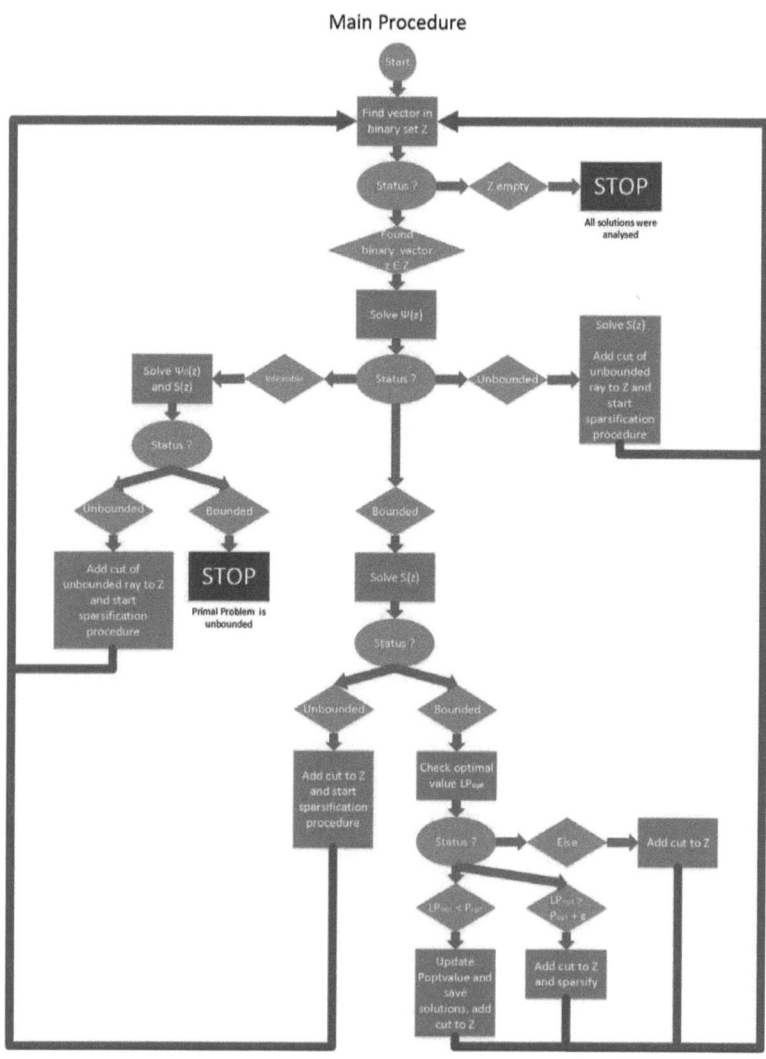

Figure 6.2: Flow chart of the main procedure.

6.5 Pseudo-Code of Stolibi

to Z where z^* is the binary vector which started the iteration. This would be done until each and every point in $\{0,1\}^m$ is analyzed. In order to avoid this kind of enumeration, the algorithm stops if an unbounded model with negative extreme rays is found.

The same cut (6.21) is added if the solution is degenerated, i.e., it holds $v_1 = 0$ and $v_2 = 0$.

6.5.3 Sparsification Procedure

In essence, the sparsification procedure is a heuristic to shorten the branching on the binary variables. The algorithm would terminate without this procedure, but, it would – in the worst case – be just a bit better than an enumeration algorithm.

The procedure starts with a ray cut or a point cut of the form

$$\sum_{\omega=1}^{N} \sum_{l \in J_1(\omega)} z_{\omega,l} + \sum_{\omega=1}^{N} \sum_{k \in J_2(\omega)} (1-z)_{\omega,k} \geq 1 \qquad (6.22)$$

and uses the fact that if subsets $K_1(\omega) \subset J_1(\omega)$ and $K_2(\omega) \subset J_2(\omega)$ can be found such that the cut

$$\sum_{\omega=1}^{N} \sum_{l \in K_1(\omega)} z_{\omega,l} + \sum_{\omega=1}^{N} \sum_{k \in K_2(\omega)} (1-z)_{\omega,k} \geq 1 \qquad (6.23)$$

is also valid, then, the new cut is tighter than the so called *root* cut (6.22) since there are less members in the sums so that it is harder to fulfill the greater-or-equal condition. In that case, the subcut (6.23) can be added to Z instead of (6.22).

A cut (6.22) will be divided into two *sparsified* cuts of the above type (6.23), with $K_1(\omega)$ being the first set of the first cut and $K_2(\omega)$ being the second set of the first cut, and $J_1(\omega) \setminus K_1(\omega)$ being the first set of the second cut whereas $J_2(\omega) \setminus K_2(\omega)$ is the second set of the second cut. So, $J_1(\omega)$ and $J_2(\omega)$ will be partitioned into two "disjunct" subcuts.

The authors Hu et al. [50] did not specify how to divide the root cut, so I chose to divide the cut in such a way that every second index will be in the same sparsified cut.

Whereas the main procedure only works on the dual problems $\text{DLP}(\theta, z)$ (and their homogeneous counterparts), this procedure works mainly with the relaxed primal problem with additional constraints and only uses the dual problem to regain feasibility in certain cases.

6. STOLIBI – AN ALGORITHM FOR LINEAR STOCHASTIC BILEVEL PROBLEMS

In order to test the new sparsified cut (6.23) for validity, the following linear problem is solved

$$\min \left\{ c^\top x + \sum_{\omega=1}^{N} d_\omega^\top y(\omega) \; : \; \begin{array}{ll} Ax = b, \, x \geq 0, & \\ Wy(\omega) = h_\omega - Tx, & \forall \omega \\ y(\omega) \geq 0, & \forall \omega \\ y_i(\omega) = 0, & i \in K_2(\omega), \forall \omega \\ W^\top u(\omega) + \mu(\omega) = q_\omega, & \forall \omega \\ \mu(\omega) \geq 0 & \forall \omega \\ \mu_i(\omega) = 0 & i \in K_1(\omega), \forall \omega \end{array} \right\} \quad (6.24)$$

This problem is built from the relaxed bilevel problem shown in the section preprocessing 6.5.1 and can also be compared to the model LP(α) (6.2). But, due to the fact that it only partially treats the missing complementarity constraint, its solution and solution value do not have to be feasible or a bound for the KKT reformulated bilevel problem. Additionally, the solution value can also be worse than the best solution because the sets $K_1(\omega)$ and $K_2(\omega)$ fix certain values in the above model. Therefore, if the solution value of problem (6.24) is (finite, but) greater than the current best bound, the sparsified cut can be added to Z since it does not produce a better bound nor an unbounded problem.

If, instead, problem (6.24) is infeasible, the same can be done. If this problem is infeasible dependent on the sets $K_1(\omega)$ and $K_2(\omega)$, it might still be feasible for other sets (i.e., binary vectors z). The main procedure will find a certificate of infeasibility at the end.

But, if the problem is bounded with better objective value (including unboundedness), it has to be checked if the complementarity conditions are fulfilled. If so, the bound P_{opt} can be updated as well as the binary vector z_{opt} with

$$z_{opt,i} = 0 \quad \text{if } y_i > 0 \quad \text{and} \quad z_{opt,i} = 1 \quad \text{if } \mu_i > 0.$$

If both variables y_i and μ_i are zero, $z_{opt,i}$ can be chosen arbitrarily. If the problem is unbounded and the complementarity conditions apply, the algorithm stops because a certificate of unboundedness has been found.

Now, the case when the complementarity conditions are not fulfilled is a bit harder. Since this cut seems promising to produce a better bound, a feasibility recovery procedure will be started.

6.5 Pseudo-Code of Stolibi

Feasibility Recovery Procedure

In order to create a feasible solution, the procedure takes the values of the leader's variable x^* and solves each lower level $Y(x^*, \omega), \omega = 1, \ldots, N$. The solution (x^*, y^*) – if existent – is most likely not optimal, but bilevel feasible. Therefore, its value

$$c^\top x^* + \sum_{\omega=1}^{N} \pi_\omega \cdot d(\omega)^\top y^*(\omega)$$

creates an upper bound to the optimal solution if the follower's problems are all bounded and feasible. In addition, a feasible binary vector can be created (similar to the above binary vector z_{opt}) with

$$z_i^* = 0 \quad \text{if } y_i^* > 0 \quad \text{and} \quad z_i^* = 1 \quad \text{else.}$$

With this binary vector, $\psi(z^*)$ and $S(z_\omega^*)$ can be solved and the same cases can be distinguished and treated as in the main procedure, except for the part where the sparsification procedure starts.

All this can be done if the follower's problems are feasible and bounded. It can also be done if all follower's problems are feasible and some are unbounded. A representative of the unbounded ray y^* can still produce the binary vector z^* and the dual problem may then find a certificate of unboundedness. In these cases, the sparsified cut that started the feasibility recovery procedure will be deleted from the sparsification procedure without adding it to Z.

The only undesirable case is when one of the follower's problems is infeasible. In that case, no new cut can be generated and the old cut cannot be used since it is not a complete vector to parameterize the dual problems. One option would be to just drop that constraint and start over with the sparsification procedure. But, some information were already gathered about that cut, so, the authors Hu et al. [50] proposed to create a set named Z_{wait} which collects these cuts and the solution value they produced on the problem (6.24). Every time the upper bound P_{opt} is lowered during the main procedure or the sparsification procedure, this set is checked on cuts that have a solution value greater, i.e., worse, than the new bound P_{opt}. These cuts will be "activated" and put into Z.

Sparsification Procedure

Data Input: $\sum_{\omega=1}^{N} \sum_{l \in J_1(\omega)} z_{\omega,l} + \sum_{\omega=1}^{N} \sum_{k \in J_2(\omega)} (1-z)_{\omega,k} \geq 1$ from main procedure

6. STOLIBI – AN ALGORITHM FOR LINEAR STOCHASTIC BILEVEL PROBLEMS

1. **Initialization:** Partition the cut $\{J_1, J_2\}$ into two cuts $\{K_1, K_2\}$ and its relative complement $\{K_1^C, K_2^C\}$

2. **While** there are cuts to treat:
   Extract one cut and solve the problem (6.24) according to the sets $K_1(\omega)$ and $K_2(\omega)$ defining the cut.

   (a) **If** problem (6.24) is infeasible: Add cut to Z, start again with 2.

   (b) **Else if** problem (6.24) is unbounded: check complementarity constraints
   If fulfilled, STOP - bilevel problem is unbounded,
   Else start feasibility recovery procedure. **If** it is successfull, add cut to Z, start again with 2.
   Else (feasibility recovery procedure is not successful) put cut into a waiting pool, start again with 2.

   (c) **Else** problem (6.24) is bounded:
   If solution value LP_{rlx} is $P_{opt} \leq LP_{rlx} < P_{opt} + EPS$, add cut to Z, start again with 2,
   Else if solution value LP_{rlx} is $LP_{rlx} \geq P_{opt} + EPS$, add cut to Z, sparsify it in two subcuts, and start again with 2,
   Else solution value is better than the current best value: Check if complementarity constraints are fulfilled.

   - **If** so, add cut to Z, update the best bound P_{opt} and the best solution, start again with 2.
   - **Else** start feasibility recovery procedure.
     If it is successfull, add cut to Z, start again with 2.
     If not, put cut into a waiting pool, start again with 2.

   End

3. Go back to main procedure.

The sparsification procedure terminates after a finite number of steps due to the fact that it starts with one cut of finite length and in the worst case sparsifies the cut in every step again until it searches through all indices of it.

6.5 Pseudo-Code of Stolibi

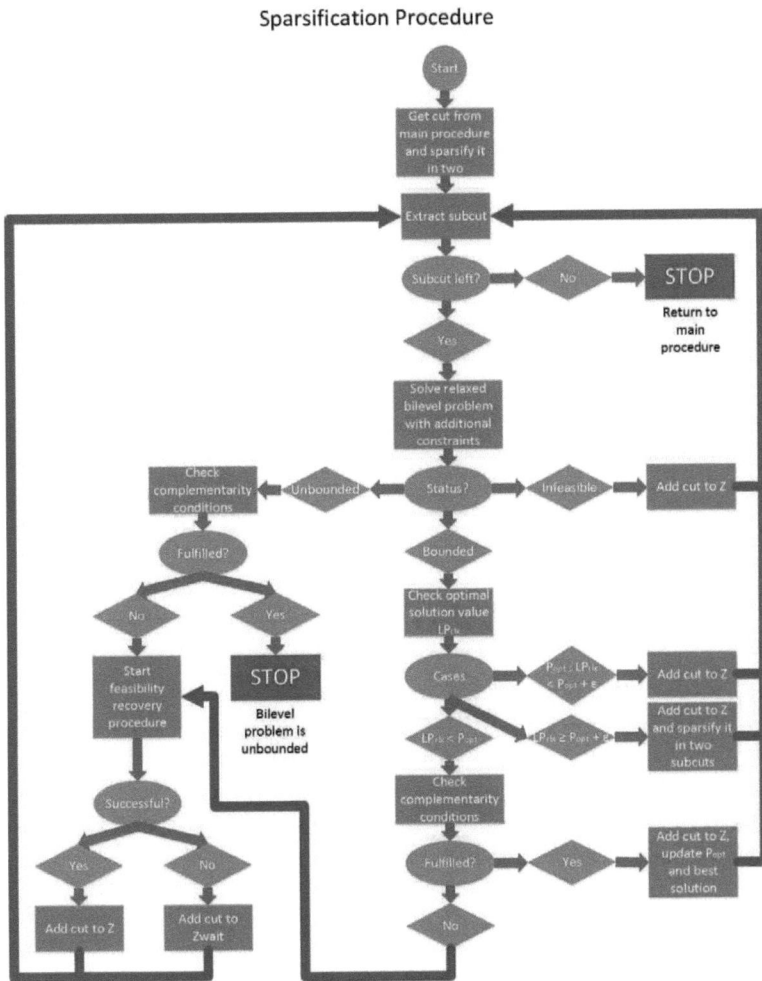

Figure 6.3: Flow chart of the sparsification procedure.

6. STOLIBI – AN ALGORITHM FOR LINEAR STOCHASTIC BILEVEL PROBLEMS

6.6 Modulations of the Original Algorithm

This section presents the modulations that have been made on the algorithm of Hu et al. [50] in order to accommodate it for the stochastic bilevel case. I will go through the algorithm procedural.

For the preprocessing, the authors proposed to use "simple cuts". These cuts would combine certain nonbasic values of a solution with those nonbasic variables into one inequality. In the stochastic case, this would result in combining several scenario variables in one inequality, thus, complicating the use of decomposition methods. Therefore, they were omitted here.

In the main procedure, the selection of $z \in Z$ was defined to be $\min z, z \in Z$. Additionally and more important, the problem $\text{DLP}(\theta, z)$ is decomposed into two subproblems of which one itself decomposes scenariowise.

Degenerated solutions, i.e., solutions with $v_1 = 0$ and $v_2 = 0$, are not treated by Hu et al. [50]. In such cases, their algorithm would not add any cut to Z anymore and not terminate. Stolibi instead will just add a cut to Z which precisely cuts off that z^* which started the iteration. The cut would be

$$\sum_{i:z_i^*=0} z_i + \sum_{j:z_j^*=1} (1-z)_j \geq 1.$$

Adding to Hu et al. [50], the feasibility recovery procedure was rethought from the very beginning and specialized to the stochastic bilevel structure.

The algorithm was implemented using the programming language C++ on a linux system (`Ubuntu/Linaro 4.6.3-1ubuntu5`, `g++ version 4.6.3`). The optimizer Gurobi [47] was used to solve all stated models.

Stolibi was implemented only on the knowledge of the paper by Hu et al. [50]. Therefore, the code and, especially, the computational approach in detail might most probably differ a lot due to the lack of information.

6.7 Proof of Correctness and Finiteness

Following the paper of Hu et al. [50], it will be shown:

Lemma 6.9
The algorithm terminates after a finite number of iterations.

6.7 Proof of Correctness and Finiteness

Proof:

The finiteness is due to the facts that:

a) The set of binary vectors Z is finite,

b) There are only finitely many cuts, and

c) Each iteration of the algorithm generates a new binary vector that is distinct from all those previously generated.

Statements a) and b) are obvious. But, compared to the work of Hu et al. [50], statement c) has to be shown a bit different since the authors missed the degenerated case when no entry of $v = (v_1, v_2)$ is greater zero. In such a case, no cut would be generated in their algorithm and the procedure would cycle. Here, the case is covered by the operation of adding a cut to Z that just cuts off the z-vector that was generated at last in the first step of the main procedure.

Otherwise, it also holds that in every iteration a new cut is generated that at least cuts off the previous found z-vector. □

The proof of this lemma was also provided in the paper of Hu et al. [50] leaving the mentioned gap regarding c) under degeneracy.

Now, I am in the position to formulate the main result of the present thesis.

Theorem 6.10
The algorithm stolibi finds an optimal solution for the optimistic linear stochastic bilevel problem.

Proof:

Since the algorithm terminates after a finite number of steps, it has to be shown that the procedures do not cut off the optimal solution without saving it (as the optimal solution).

First of all, any z-vector that produces an improved finite solution value is saved. Thus, if (one of) the optimal z-vector(s) is generated in the first step of the main procedure, it is also saved as the best solution and would only be overwritten afterwards by vectors with better solution value.

So, it has to be shown that the cuts generated in the main and sparsification procedures do not accidentally cut off the best solution. Therefore, it is helpful to list the times whenever a cut is added and which kind of cut it is.

6. STOLIBI – AN ALGORITHM FOR LINEAR STOCHASTIC BILEVEL PROBLEMS

The cuts in the main procedure are added if:

1. $\psi(z)$ or at least one $S(z_\omega)$ are unbounded (ray cut),

2. $\psi(z)$ is infeasible and $\psi_{homo}(z)$ or $S(z_\omega)$ is unbounded (ray cut),

3. $\psi(z)$ as well as all $S(z_\omega)$ are bounded (point cut).

The validation of these cuts was already shown at the end of section 6.4. More interesting are the cuts that are produced in the sparsification procedure when:

1. the relaxed primal problem with additional constraints (6.24) is infeasible,

2. the relaxed primal problem with additional constraints (6.24) is bounded and the solution value is worse than the currently best solution,

3. the relaxed primal problem with additional constraints (6.24) is bounded, the solution value is better than the currently best solution, and the complementarity conditions are fulfilled (including the unbounded case),

4. the cut is produced in the feasibility recovery procedure.

For the first three cases, it is important to mention that the relaxed primal problem with additional constraints (6.24) is equivalent to solving the parameterized linear problem

$$\min \left\{ c^\top x + \sum_{\omega=1}^{N} d_\omega^\top y(\omega) : \begin{array}{ll} Ax = b, \, x \geq 0, & \\ Wy(\omega) = h_\omega - Tx, & \forall \omega \\ y(\omega) \geq 0, & \forall \omega \\ W^\top u(\omega) + \mu(\omega) = q_\omega, & \forall \omega \\ \mu(\omega) \geq 0 & \forall \omega \\ \mu_i(\omega) \leq \theta z^*_{\omega,i} & i \in K_1(\omega), \forall \omega \\ y_i(\omega) \leq \theta(1 - z^*_{\omega,i}), & i \in K_2(\omega), \forall \omega \end{array} \right\} \quad (6.25)$$

where $z^*_{\omega,i} = 0$ for $i \in K_1(\omega)$ and $z^*_{\omega,i} = 1$ for $i \in K_2(\omega)$ (and the rest of the binary vector z^* can be set arbitrarily to zero or one). Thus, z^* is exactly the vector violating

$$\sum_{\omega=1}^{N} \sum_{l \in K_1(\omega)} z_{\omega,l} + \sum_{\omega=1}^{N} \sum_{k \in K_2(\omega)} (1 - z)_{\omega,k} \geq 1, \quad (6.26)$$

the cut that is treated in the sparsification procedure at that moment.

So, if the relaxed primal problem with additional constraints (6.24) is infeasible, the cut can be added to Z because only if all binary vectors produce infeasible problems, the bilevel problem is infeasible. This cut will allow to find such a certificate faster.

In the second case, when the problem is bounded and the solution value is worse, i.e., greater, than the current best bound P_{opt}, the cut can also be added to Z. The reason for doing so is that any completion of that vector z^* would only produce worse optimal solution values or infeasible problems.

In the third case, the algorithm found a bilevel feasible solution with better solution value. The latter is saved as P_{opt} and z^* is complemented according to be saved as z_{opt}. Therefore, the subcut (6.26) can also be added to Z.

If the feasibility recovery procedure has to start, the solution value of (6.25) was better than P_{opt}, but the complementarity conditions were not all fulfilled. During that procedure, a cut is only added to Z if all follower's problems are feasible for the values of x^* set as parameters. If so, the subcut (6.26) is disregarded and a new binary vector is constructed from the solution y'. If not, the cut is added to Z_{wait}; this will be treated in the next paragraph. In the former case, the same practice as in the main procedure is done with the new binary vector. These cuts were already shown to be valid.

At last, a cut is added to Z when P_{opt} is lowered and a cut saved in Z_{wait} had a solution value now worse than P_{opt}. This is also valid due to the same reason as in the second case of the sparsification procedure.

Additionally, the certificates for unboundedness and infeasibility were also already shown to be correct in section 6.3. Therefore, the algorithm works correctly. □

6.8 Requirements and Input

Actually, there are very few conditions on the problems. The algorithm will find infeasibility or unboundedness. The only restrictions are made by the form of the input data. The bilevel problem has to be an equality problem, meaning that every constraint (except the nonnegativity constraints) has to hold by equality. This can be easily achieved with slack variables if the problem is actually dependent on inequalities.

The model has to be a minimization problem. This can also be easily accomplished by taking the negative of the objective function. Additionally, it is intrinsically assumed that the lower level problem is a minimization problem. This will not be checked by

6. STOLIBI – AN ALGORITHM FOR LINEAR STOCHASTIC BILEVEL PROBLEMS

the procedure while the leader's model is checked on that. The algorithm only works on problems with at least two scenarios.

Important for the usage of the algorithm on other PCs is the installation of the Optimizer Gurobi [47] and the adjustment of the makefile if the libraries of gurobi are not put in the standard place. As a research code, it was implemented and checked on a linux ubuntu distribution, exclusively.

The input has to be done through two files. Details were already given in section 6.5.1 describing the preprocessing. It is important to remember that the stochastics only influence the follower's cost vector q, the follower's right-hand side h, and, consequently, the leader's cost vector for the follower's variables. The remaining data (i.e., the matrices W and T) are just copied for every scenario starting with the second.

Finally, the beginning of some of the vectors' names are set. The variables have to begin with "X" or "Y", the upper level (resp. lower level) follower's cost vector has to begin with "D" (resp. "Q") and the right-hand side has to be named "changeh" beginning with the second scenario.

6.9 Computational Results

In order to check the correctness of stolibi, I also implemented an algorithm in C++ which uses the same data input as stolibi, described in sections 6.8 and 6.5.1. The data is then merged to the following mixed-integer linear problem

$$\min \left\{ c^\top x + \sum_{\omega=1}^{N} \pi(\omega) d_\omega^\top y(\omega) \ : \ \begin{array}{l} Ax = b, x \geq 0, \\ Wy(\omega) = h_\omega - Tx, \quad \forall \omega \\ y(\omega) \geq 0, \quad \forall \omega \\ W^\top u(\omega) + \mu(\omega) = q_\omega, \quad \forall \omega \\ \mu(\omega) \geq 0, \quad \forall \omega \\ \mu(\omega) \leq Mz(\omega), \quad \forall \omega \\ y(\omega) \leq M(1 - z(\omega)), \quad \forall \omega \\ z(\omega) \in \{0,1\}^m, \quad \forall \omega \end{array} \right\} \quad (6.27)$$

The algorithm solves the KKT reformulated bilevel problem in the standard approach, optimizing the above model for one M set to a certain high value (e.g. $M = 1000$ for smaller problems is reasonable).

6.9 Computational Results

In most cases, the problem has to be solved for different values of M in order to be sure that the problem is not unbounded or that an undersized M surpresses the correct solution. Still, if one can be sure that the lower level problems are bounded and M is set appropriate, it is safe to assume that the solution of the algorithm is correct.

As for stolibi, the same Optimizer Gurobi [47] was used to solve this problem. Table 6.1 shows the performance of stolibi for certain artificial examples. Tables 6.2 and 6.3 show the results of two stochastic bilevel problems with 10 scenarios compared to the solutions when perfect information is available, i.e., when the leader knows which scenario will be realized.

SCEN	SIZE of y	Runtime	Solution Value	Solution found in
3	5	0.07 sec	Unbd	Main Procedure
4	6	0.4 sec	10.0	Feasibility rec. proc.
10	5	2.4 sec	-144.0	Feasibility rec. proc.
10	20	0.03 sec	-163.5	Preprocessing
5	5	0.4 sec	100.9	Feasibility rec. proc.
6	5	4.1 sec	124.75	Feasibility rec. proc.
7	5	37.4 sec	157.5	Feasibility rec. proc.
8	5	3 min 7.2 sec	175.25	Feasibility rec. proc.
9	5	3 min 12.0 sec	215.0	Feasibility rec. proc.
10	5	80 min 4.7 sec	232.25	Feasibility rec. proc.
20	5	> 4 h (stopped)	(483.75)	(Feasibility rec. proc.)

Table 6.1: Performance of stolibi

SCEN represents the number of scenarios in that model and "SIZE of y" indicates the dimension of each scenario's lower level decision vector. The first block shows the results for some arbitrary problems, whereas, the second block represents the same problem with a growing number of scenarios.

It is pretty interesting that, in most of the cases, the solution was found in the first iteration of the main procedure when the S-models where unbounded and the sparsification procedure is started with that ray cut. One can see that the algorithm works very fast for smaller problems. The correctness of the solutions was verified by the above alternative.

6. STOLIBI – AN ALGORITHM FOR LINEAR STOCHASTIC BILEVEL PROBLEMS

However, the second block shows that with a growing number of scenarios, the algorithm becomes much slower. Still, the solutions for higher numbers of scenarios were also found the first time the sparsification procedure was started. This could be used for later research or in the use of a heuristic.

The high running time for comparatively small examples, e.g. with scenario size 10 or greater, can be explained by the size of the binary problem Z and the need to normally add a high number of cuts until it becomes infeasible. Besides, the addition of many cuts causes the problem to be solvable even harder, thus, longer.

Scenario	Objective Value	X values	Obj. value with $\bar{x}$
1	-25.0	(1, 2.0, 3, 4, 5)	-25.0
2	-33.6	(1, 1.6, 3, 4, 5)	5
3	-25.0	(1, 2.0, 3, 4, 5)	-25.0
4	-35.0	(1, 2.0, 3, 4, 5)	-35.0
5	35.0	(1, 2.0, 3, 4, 5)	35.0
6	-5.0	(1, 2.0, 3, 4, 5)	-5.0
7	20.0	(0, 2.0, 3, 4, 5)	35
8	-26.3	(1, 1.6, 3, 4, 5)	infeasible
9	-35.0	(1, 2.0, 3, 4, 5)	-35.0
10	6.3	(1, 1.6, 3, 4, 5)	infeasible
$\mathbb{E}$	12.4	$\bar{x} = $ (1, 2.0, 3, 4, 5)	-
stoch	-2.0	(0, 2.0, 3, 4, 5)	$EEV = -$

Table 6.2: Comparison of the perfect information solutions and the expected value solution for example 1.

In addition to that, it might be interesting how much influence the uncertainty has on the solution and its value. Two examples will provide more insight (see tables 6.2 and 6.3), each with a five-dimensional leader's variable, ten scenarios, and a five-dimensional follower's decision vector for each scenario. All problems are minimization problems as described in section 6.8. In the stochastic models, each scenario had the same probability, i.e., $\pi_\omega = \frac{1}{10}$.

The stochastic problem is solved with stolibi and the solution can be found under the name "stoch" in the last row. The first ten rows of the tables show the results when each scenario is solved as an individual deterministic bilevel problem. Thus, these problems

6.9 Computational Results

give the solutions if the leader had perfect information about the follower's problem. The remaining row, symbolized with $\mathbb{E}$, depicts the solution if the data exposed to randomness is each replaced by its expected value.

In the first table 6.2, one can see that the objective values vary a lot between +35 and −35. More importantly, scenarios 8 and 10 would not be feasible if the solution of the expected value problem was used. This implies that the stochastic approach is very appropriate at this place.

In addition, the value of the stochastic solution (VSS) and the expected value of perfect information (EVPI) – see Chapter 4 in Birge and Louveaux [15] – can be examined.

For table 6.2, the VSS cannot be determined because it depends on the EEV (the EEV is the expected result of using the expected value solution, i.e., the weighted sum of the last column), but

$$EVPI = stoch - WS = -2 - (-11.32) = 9.32$$

where WS is the expectation of the optimal values of all possible scenarios, i.e., $WS = \frac{1}{10}\sum_{i=1}^{10}$ objective value of deterministic version of scenario i. The value shows that the

Scenario	Objective Value	X values	Obj. value with $\bar{x}$
1	-132	(1, 2, 3, 4, 0)	-132
2	6	(1, 0, 3, 0, 5)	68
3	-417	(0, 0, 3, 4, 0)	-412
4	-190	(1, 2, 3, 4, 0)	-162
5	-262	(1, 0, 3, 4, 0)	-262
6	-232	(1, 0, 3, 4, 0)	-232
7	-4	(0, 0, 3, 0, 5)	128
8	-140	(1, 2, 3, 4, 0)	-72
9	-442	(1, 0, 3, 4, 0)	-442
10	-59	(1, 0, 3, 0, 5)	68
$\mathbb{E}$	-228.8	$\bar{x} = (1, 0, 3, 4, 0)$	-
stoch	-144	(1, 0, 3, 4, 0)	$EEV = $ -144

Table 6.3: Comparison of the perfect information solutions and the expected value solution for example 2.

6. STOLIBI – AN ALGORITHM FOR LINEAR STOCHASTIC BILEVEL PROBLEMS

knowledge of future events should not cost more than 9.32.

Table 6.3 instead shows that even though the objective values differ a lot between -442 and $+6$, the solution of the stochastic problem and the solution of the expected value problem $\mathbb{E}$ are the same. Still, the evaluation of the stochastic solution can be done through

$$VSS = EEV - stoch = -144 + 144 = 0$$

which is clear since the solutions of $\mathbb{E}$ and $stoch$ are the same. But, the expected value of perfect information

$$EVPI = stoch - WS = -144 - (-214.48) = 70.48$$

shows that it would be much better to know the future event if possible. In other words, this problem's scenariowise solution values have a rather high variability compared to the stochastic solution.

7
Conclusion

This thesis is concerned with linear stochastic bilevel problems and how decomposition methods known from stochastic programming can be used here. An overview on the literature in the field of stochastic bilevel problems was given. Two decomposition methods were reviewed.

Few properties of the (multi-)functions appearing in linear bilevel problems were examined, as well as their reformulations, and random parameters were introduced.

Two approaches have been made in order to apply decomposition methods, one using the optimal value function, the other using the KKT conditions. The first approach via the value function led to the (preliminary) outcome that the reverse convex constraints it produces defy incorporation into a cutting-plane decomposition scheme of, say, Benders type. To the author's knowledge this remains an open problem in the research literature.

Capturing lower-level optimality via the KKT conditions, however, the problem can be reformulated following the big-M idea without exercising it numerically. Controlling the combinatorics behind the complementarity conditions by Boolean entities understood as parameters rather than variables, yields a reformulation of the original problem as a linear program with integer parameters. In this setting, for fixed parameters, dualization becomes feasible which leads to a model that is amenable to decomposition into single-scenario problems. Embedding into a branch-and-bound scheme over the Boolean parameters results in a scenario decomposition algorithm for linear stochastic bilevel programs.

7. CONCLUSION

The algorithm was analyzed, implemented, and evaluated. As a research code, its purpose was to provide a proof-of-concept rather than outperforming commercial general-purpose solvers. Still, it shows some advantages that might be of use in future research. The structure of the dual problem allowed for decomposition, but only parts of it were implemented. The Dantzig-Wolfe decomposition of one of the subproblems was left out and might become a run-time saver for bigger problems if implemented accordingly. In addition, the implementation was done "my way" and without the usage of parallelization. The latter might win some time, e.g. when the S-models are solved simultaneously, or the problems occuring in the Dantzig-Wolfe decomposition could be solved at the same time.

Therefore, this thesis lays the foundation to further research on decomposition methods in that direction.

References

[1] E. AIYOSHI AND K. SHIMIZU. **A solution method for the static constrained Stackelberg problem via penalty method.** *Automatic Control, IEEE Transactions on*, 29(12):1111–1114, 1984.

[2] S. M. ALIZADEH, P. MARCOTTE, AND G. SAVARD. **Two-stage stochastic bilevel programming over a transportation network.** *Transportation Research Part B: Methodological*, 58:92–105, 2013.

[3] K. J. ARROW, L. HURWICZ, AND H. UZAWA. **Constraint qualifications in maximization problems.** *Naval Research Logistics Quarterly*, 8(2):175–191, 1961.

[4] J.-A. AUDESTAD, A. A. GAIVORONSKI, AND A. WERNER. **Extending the stochastic programming framework for the modeling of several decision makers: pricing and competition in the telecommunication sector.** *Annals of Operations Research*, 142(1):19–39, 2006.

[5] B. BANK, J. GUDDAT, D. KLATTE, B. KUMMER, AND K. TAMMER. *Non-linear Parametric Optimization.* Birkhäuser Basel, 1983.

[6] J. F. BARD. **Optimality Conditions for the Bilevel Programming Problem.** *Naval research logistics quarterly*, 31(1):13–26, 1984.

[7] J. F. BARD. *Practical Bilevel Optimization: Algorithms and Applications.* Kluwer Academic Publishers, 1998.

[8] J. F. BARD AND J. E. FALK. **An explicit solution to the multi-level programming problem.** *Computers & Operations Research*, 9(1):77–100, 1982.

[9] J. F. BENDERS. **Partitioning procedures for solving mixed-variables programming problems.** *Numerische mathematik*, 4:238–252, 1962.

[10] B. BEREANU. **Minimum risk criterion in stochastic optimization.** *Economic Computation and Economic Cybernetics Studies and Research*, 2:31–39, 1981.

[11] D. P. BERTSIMAS AND J. N. TSITSIKLIS. *Introduction to Linear Optimization.* Athena Scientific Series in Optimization and Neural Computation, 6. Athena Scientific, 1997.

[12] W. BIALAS AND M. KARWAN. **Two-level linear programming.** *Management Science*, 30:1004–1020, 1984.

[13] W. BIALAS, M. KARWAN, AND J. SHAW. **A parametric complementary pivot approach for two-level linear programming.** State University of New York at Buffalo, 1980.

[14] S. I. BIRBIL, G. GÜRKAN, AND O. LISTEŞ. **Simulation-based solution of stochastic mathematical programs with complementarity constraints: Sample-path analysis.** Technical report, ERIM Report Series Research in Management, 2004.

[15] J. R. BIRGE AND F. LOUVEAUX. *Introduction to stochastic programming.* Springer, 1997.

[16] J. BRACKEN AND J. MCGILL. **Mathematical programs with optimization problems in the constraints.** *Operations Research*, 21:37–44, 1973.

[17] J. V. BURKE AND M. C. FERRIS. **Weak sharp minima in mathematical programming.** *SIAM Journal on Control and Optimization*, 31(5):1340–1359, 1993.

[18] W. CANDLER AND R. NORTON. **Multilevel programming.** Technical report, World Bank Developement Research Center, Washington D.C., 1977.

[19] W. CANDLER AND R. TOWNSLEY. **A linear two-level programming problem.** *Computers & Operations Research*, 9(1):59–76, 1982.

[20] M. CARRIÓN, J. M. ARROYO, AND A. J. CONEJO. **A bilevel stochastic programming approach for retailer futures market trading.** *Power Systems, IEEE Transactions on*, 24(3):1446–1456, 2009.

[21] L. M. CASE. *An l1 Penalty Function Approach to the Nonlinear Bilevel Programming Problem.* PhD thesis, University of Waterloo, Ontario, Canada, 1999.

[22] Y CENSOR AND S. A. ZENIOS. *Parallel optimization: Theory, algorithms, and applications.* Oxford University Press, 1997.

[23] Y. CHEN AND M. FLORIAN. **The nonlinear bilevel programming problem: Formulations, regularity and optimality conditions.** *Optimization*, 32(3):193–209, 1995.

[24] V. CHVÁTAL. *Linear Programming.* WH Freeman and Company, New York, 1983.

[25] B. COLSON, P. MARCOTTE, AND G. SAVARD. **A trust-region method for nonlinear bilevel programming: algorithm and computational experience.** *Computational Optimization and Applications*, 30(3):211–227, 2005.

[26] B. COLSON, P. MARCOTTE, AND G. SAVARD. **An overview of bilevel optimization.** *Annals of Operations Research*, 153:235–256, 2007.

[27] G. B. DANTZIG AND P. WOLFE. **The decomposition principle for linear programs.** *Operations Research*, 8:101–111, 1960.

[28] S. DEMPE. *Foundations of Bilevel Programming.* Kluwer Academic Publishers, 2002.

REFERENCES

[29] S. DEMPE. **Annotated Bibliography on Bilevel Programming and Mathematical Programs with Equilibrium Constraints**. *Optimization: A Journal of Mathematical Programming and Operations Research*, 52:333–359, 2003.

[30] S. DEMPE AND J. F. BARD. **Bundle trust-region algorithm for bilinear bilevel programming**. *Journal of Optimization Theory and Applications*, 110(2):265–288, 2001.

[31] S. DEMPE AND J. DUTTA. **Is bilevel programming a special case of a mathematical program with complementarity constraints?** *Mathematical Programming*, 131(1-2):37–48, 2012.

[32] S. DEMPE, J. DUTTA, AND B. S. MORDUKHOVICH. **New necessary optimality conditions in optimistic bilevel programming**. *Optimization*, 56(5-6):577–604, 2007.

[33] S. DEMPE, V. V. KALASHNIKOV, AND N. KALASHNYKOVA. **Optimality conditions for bilevel programming problems**. In *Optimization with Multivalued Mappings*, pages 3–28. Springer, 2006.

[34] S. DEMPE, V. V. KALASHNIKOV, G. A. PÉREZ-VALDÉS, AND N. I. KALASHNYKOVA. **Natural gas bilevel cash-out problem: Convergence of a penalty function method**. *European Journal of Operational Research*, 215(3):532–538, 2011.

[35] S. DEMPE AND A. B. ZEMKOHO. **The Generalized Mangasarian-Fromowitz Constraint Qualification and Optimality Conditions for Bilevel Programs**. *J. Optimization Theory and Applications*, 148(1):46–68, 2011.

[36] S. DEMPE AND A. B. ZEMKOHO. **On the Karush–Kuhn–Tucker reformulation of the bilevel optimization problem**. *Nonlinear Analysis: Theory, Methods & Applications*, 75(3):1202–1218, 2012.

[37] S. DEMPE AND A. B. ZEMKOHO. **The bilevel programming problem: reformulations, constraint qualifications and optimality conditions**. *Mathematical Programming*, pages 1–27, 2013.

[38] M. DYER AND L. STOUGIE. **Computational complexity of stochastic programming problems**. *Mathematical Programming*, 106(3):423–432, 2006.

[39] J. E. FALK AND R. M. SOLAND. **An algorithm for separable nonconvex programming problems**. *Management Science*, 15(9):550–569, 1969.

[40] M. FAMPA, L. A. BARROSO, D. CANDAL, AND L. SIMONETTI. **Bilevel optimization applied to strategic pricing in competitive electricity markets**. *Computational Optimization and Applications*, 39(2):121–142, 2008.

[41] M. L. FLEGEL. **Constraint qualifications and stationarity concepts for mathematical programs with equilibrium constraints**. PhD thesis, 2005.

[42] J. FORTUNY-AMAT AND B. MCCARL. **A representation and economic interpretation of a two-level programming problem**. *Journal of the operational Research Society*, pages 783–792, 1981.

[43] R. M. FREUND. **Postoptimal analysis of a linear program under simultaneous changes in matrix coefficients**. In *Mathematical Programming Essays in Honor of George B. Dantzig Part I*, pages 1–13. Springer, 1985.

[44] S. A. GABRIEL AND F. U. LEUTHOLD. **Solving discretely-constrained MPEC problems with applications in electric power markets**. *Energy Economics*, 32(1):3–14, 2010.

[45] A. A. GAIVORONSKI AND A. WERNER. **Stochastic Programming Perspective on the Agency Problems Under Uncertainty**. In *Managing Safety of Heterogeneous Systems*, pages 137–167. Springer, 2012.

[46] R. GOLLMER. **On linear multiparametric optimization with parameter-dependent constraint matrix**. *Optimization*, 16(1):15–23, 1985.

[47] GUROBI OPTIMIZATION INC. **Gurobi Optimizer Reference Manual**, 2014. Available from: http://www.gurobi.com.

[48] P. HANSEN, B. JAUMARD, AND G. SAVARD. **New branch-and-bound rules for linear bilevel programming**. *SIAM Journal on Scientific and Statiscal Programming*, 13:1194–1217, 1992.

[49] R. HENRION AND T. SUROWIEC. **On calmness conditions in convex bilevel programming**. *Applicable Analysis*, 90(6):951–970, 2011.

[50] J. HU, J. E. MITCHELL, J.-S. PANG, K. P. BENNETT, AND G. KUNAPULI. **On the Global Solution of Linear Programs with Linear Complementarity Constraints**. *SIAM J. Optim.*, 19(1):445–471, 2008.

[51] J. HU, J. E. MITCHELL, J.-S. PANG, AND B. YU. **On linear programs with linear complementarity constraints**. *Journal of Global Optimization*, 53(1):29–51, 2012.

[52] S. HULSURKAR, M. P. BISWAL, AND S. B. SINHA. **Fuzzy programming approach to multi-objective stochastic linear programming problems**. *Fuzzy Sets and Systems*, 88(2):173–181, 1997.

[53] IBM. **ILOG CPLEX Optimization Studio**, 2014. Available from: http://www.ilog.com/products/cplex.

[54] Y. ISHIZUKA AND E. AIYOSHI. **Double penalty method for bilevel optimization problems**. *Annals of Operations Research*, 34(1):73–88, 1992.

[55] S. V. IVANOV. **Bilevel stochastic linear programming problems with quantile criterion**. *Automation and Remote Control*, 75(1):107–118, 2014.

[56] V. V. KALASHNIKOV, G. A. PÉREZ-VALDÉS, A. TOMASGARD, AND N. I. KALASHNYKOV. **Natural gas cash-out problem: Bilevel stochastic optimization approach**. *European Journal of Operational Research*, 206(1):18–33, 2010.

[57] P. KALL AND S. W. WALLACE. **Stochastic programming**. John Wiley and Sons Ltd, Chichester, 1994.

[58] C. D. KOLSTAD AND L. S. LASDON. **Derivative evaluation and computational experience with large bilevel mathematical programs**. *Journal of optimization theory and applications*, 65(3):485–499, 1990.

REFERENCES

[59] S. KOSUCH, P. LE BODIC, J. LEUNG, AND A. LISSER. On a stochastic bilevel programming problem. *Networks*, 59(1):107–116, 2012.

[60] R. M. KOVACEVIC AND G. C. PFLUG. Electricity swing option pricing by stochastic bilevel optimization: a survey and new approaches. *European Journal of Operational Research*, 2013.

[61] D.-Y. LIN, A. KAROONSOONTAWONG, AND S. T. WALLER. A Dantzig-Wolfe decomposition based heuristic scheme for bi-level dynamic network design problem. *Networks and Spatial Economics*, 11(1):101–126, 2011.

[62] G.-H. LIN AND M. FUKUSHIMA. Stochastic equilibrium problems and stochastic mathematical programs with equilibrium constraints: a survey. *Pacific Journal of Optimization*, 6(3):455–482, 2010.

[63] Y. LIU, H. XU, AND G.-H. LIN. Stability Analysis of Two-Stage Stochastic Mathematical Programs with Complementarity Constraints via NLP Regularization. *SIAM Journal on Optimization*, 21(3):669–705, 2011.

[64] C. M. MACAL AND A. P. HURTER. Dependence of bilevel mathematical programs on irrelevant constraints. *Computers & Operations Research*, 24(12):1129–1140, 1997.

[65] B. S. MORDUKHOVICH. Variational analysis and generalized differentiation, I: Basic theory, II: Applications. *Comprehensive Studies in Mathematics*, 330, 2006.

[66] B. S. MORDUKHOVICH AND N. M. NAM. Variational stability and marginal functions via generalized differentiation. *Mathematics of Operations Research*, 30(4):800–816, 2005.

[67] F. NOŽIČKA, J. GUDDAT, H. HOLLATZ, AND B. BANK. *Theorie der linearen parametrischen Optimierung*, 24. Akademie-Verlag, 1974.

[68] M. S. OSMAN, M. A. ABO-SINNA, A. H. AMER, AND O. E. EMAM. A multi-level non-linear multi-objective decision-making under fuzziness. *Applied Mathematics and Computation*, 153(1):239–252, 2004.

[69] F. A. PARRAGA. *Hierarchical programming and applications to economic policy*. PhD thesis, University of Arizona, 1981.

[70] M. PATRIKSSON. On the applicability and solution of bilevel optimization models in transportation science: A study on the existence, stability and computation of optimal solutions to stochastic mathematical programs with equilibrium constraints. *Transportation Research Part B: Methodological 42*, 10:843–860, 2008.

[71] M. PATRIKSSON AND L. WYNTER. Stochastic mathematical programs with equilibrium constraints. *Operations Research Letters* 25, pages 159–167, 1999.

[72] G. C. PFLUG. Some remarks on the value-at-risk and the conditional value-at-risk. *Nonconvex optimization and its applications*, 49:272–281, 2000.

[73] G. C. PFLUG AND W. RÖMISCH. *Modeling, measuring and managing risk*. World Scientific, 2007.

[74] P. PISCIELLA. *Methods for Evaluation of Business Models for Provision of Advanced Mobile Services under Uncertainty*. PhD thesis, Norwegian University of Science and Technology, Norway, 2012.

[75] A. PRÉKOPA. *Stochastic programming*. Springer, 1995.

[76] R. T. ROCKAFELLAR AND S. URYASEV. Optimization of conditional value-at-risk. *Journal of risk*, 2:21–42, 2000.

[77] E. ROGHANIAN, S. J. SADJADI, AND M.-B. ARYANEZHAD. A probabilistic bi-level linear multi-objective programming problem to supply chain planning. *Applied Mathematics and Computation*, 188(1):786–800, 2007.

[78] C. RUIZ AND A. J. CONEJO. Pool strategy of a producer with endogenous formation of locational marginal prices. *IEEE Transactions on Power Systems*, 24:1855–1866, 2009.

[79] G. SAVARD AND J. GAUVIN. The steepest descent direction for the nonlinear bilevel programming problem. *Operations Research Letters*, 15(5):265–272, 1994.

[80] A. SCHRIJVER. *Theory of Linear and Integer Programming*. Wiley, 1998.

[81] R. SCHULTZ. Stochastic programming with integer variables. *Mathematical Programming*, 97(1-2):285–309, 2003.

[82] H. D. SHERALI, A. L. SOYSTER, AND F. H. MURPHY. Stackelberg-Nash-Cournot Equilibria: Characterizations and Computations. *Operations Research*, 31(2), 1983.

[83] S. SIDDIQUI AND S. A. GABRIEL. An SOS1-based approach for solving MPECs with a natural gas market application. *Networks and Spatial Economics*, pages 1–23, 2012.

[84] M. SIMAAN. Stackelberg optimization of two-level systems. *IEEE Transactions on Systems, Man, and Cybernetics*, 1977.

[85] H. TUY, A. MIGDALAS, AND P. VÄRBRAND. A quasiconcave minimization method for solving linear two-level programs. *Journal of Global Optimization*, 4(3):243–263, 1994.

[86] M. H. VAN DER VLERK. Stochastic Programming Bibliography. World Wide Web, http://www.eco.rug.nl/mally/spbib.html, 1996-2007.

[87] R. M. VAN SLYKE AND R. WETS. L-shaped linear programs with applications to optimal control and stochastic programming. *SIAM Journal on Applied Mathematics*, 17:638–663, 1969.

[88] L. VICENTE, G. SAVARD, AND J. JÚDICE. Descent approaches for quadratic bilevel programming. *Journal of Optimization Theory and Applications*, 81(2):379–399, 1994.

[89] L. N. VICENTE AND P. H. CALAMAI. Bilevel and Multilevel Programming: A Bibliography Review. *Journal of Global Optimization*, pages 291–306, 1994.

REFERENCES

[90] H. VON STACKELBERG. *Marktform und Gleichgewicht*. Springer, 1934.

[91] U.-P. WEN AND S.-T. HSU. **Linear Bi-level Programming Problems – A Review**. *Journal of the Operations Research Society*, **42**:125–133, 1991.

[92] A. WERNER. *Bilevel stochastic programming problems: Analysis and application to telecommunications*. PhD thesis, Norwegian University of Science and Technology, 2005.

[93] H. XU AND J. J. YE. **Necessary optimality conditions for two-stage stochastic programs with equilibrium constraints**. *SIAM Journal on Optimization*, **20**(4):1685–1715, 2010.

[94] J. J. YE. **Nondifferentiable multiplier rules for optimization and bilevel optimization problems**. *SIAM Journal on Optimization*, **15**(1):252–274, 2004.

[95] J. J. YE AND D. L. ZHU. **Optimality Conditions for Bilevel Programming Problems**. *Optimization*, **33**(1):9–27, 1995.

[96] J. J. YE AND D. L. ZHU. **A Note on Optimality Conditions for Bilevel Programming Problems**. *Optimization*, **39**(4):361–366, 1997.

[97] J. J. YE AND D. L. ZHU. **New necessary optimality conditions for bilevel programs by combining the MPEC and value function approaches**. *SIAM Journal on Optimization*, **20**(4):1885–1905, 2010.

[98] J. J. YE, D. L. ZHU, AND Q. J. ZHU. **Exact penalization and necessary optimality conditions for generalized bilevel programming problems**. *SIAM Journal on Optimization*, **7**(2):481–507, 1997.

[99] D. ZHANG, H. XU, AND Y. WU. **A two stage stochastic equilibrium model for electricity markets with two way contracts**. *Mathematical Methods of Operations Research*, **71**(1):1–45, 2010.

[100] ZUSE-INSTITUT BERLIN (ZIB). **SCIP Optimization Suite**, 2014. Available from: http://scip.zib.de.

I want morebooks!

Buy your books fast and straightforward online - at one of the world's fastest growing online book stores! Environmentally sound due to Print-on-Demand technologies.

Buy your books online at
www.get-morebooks.com

Kaufen Sie Ihre Bücher schnell und unkompliziert online – auf einer der am schnellsten wachsenden Buchhandelsplattformen weltweit!
Dank Print-On-Demand umwelt- und ressourcenschonend produziert.

Bücher schneller online kaufen
www.morebooks.de

OmniScriptum Marketing DEU GmbH
Heinrich-Böcking-Str. 6-8
D - 66121 Saarbrücken
Telefax: +49 681 93 81 567-9

info@omniscriptum.com
www.omniscriptum.com

Printed by Books on Demand GmbH, Norderstedt / Germany